Investigation of Molecular Structure

SPECTROSCOPIC AND DIFFRACTION METHODS

B. C. Gilbert MA, DPhil
Department of Chemistry, University of York
Diagrams by Tony Mould

Mills & Boon Limited, London

First published in Great Britain 1975
by Mills & Boon Limited,
17–19 Foley Street, London W1A 1DR

© B. C. Gilbert 1975

ISBN 0 263 05818 2

Printed photolitho in Great Britain by
Ebenezer Baylis and Son Limited,
The Trinity Press,
Worcester, and London.

Investigation of Molecular Structure
SPECTROSCOPIC AND DIFFRACTION METHODS

Richard.

11/10/83.

Leynes college.

83 068450 4

Modern Chemistry Series

Under the supervisory editorship of D. J. Waddington, BSc, ARCS, DIC, PhD, Senior Lecturer in Chemistry, University of York, this series is specially designed to meet the demands of the new syllabuses for sixth form, introductory degree and technical college courses. It consists of self-contained major texts in the three principal divisions of the subject, supplemented by short readers and practical books in certain areas.

Major Texts

Modern Inorganic Chemistry
G. F. Liptrot MA, PhD

Modern Organic Chemistry
R. O. C. Norman MA, DSc, C Chem, FRIC, and
D. J. Waddington BSc, ARCS, DIC, PhD

Modern Physical Chemistry
Miss Judith M. Phillips BSc
J. J. Thompson MA, ARIC, and
G. R. Walker BA, BSc, ARIC,

Readers

The Periodic Table
J. S. F. Pode MA, BSc

A Mechanistic Introduction to Organic Chemistry
Glyn James MA

Investigation of Molecular Structure
B. C. Gilbert MA, DPhil

Practical Books

Organic Chemistry Through Experiment
D. J. Waddington BSc, ARCS, DIC, PhD and
H. S. Finlay BSc

Inorganic Chemistry Through Experiment
G. F. Liptrot MA, PhD

Contents

Acknowledgements

I thank Miss Kin Mya Mya, Dr. A. J. G. Crawshaw and Mr. C. J. Hall, all of the University of York, for their assistance in recording most of the spectra shown in the book. It is also a pleasure to acknowledge the sources of the following Figures and for permission to use them: Figures 1.3 (A.E.I. Ltd.), 6.15 (The Royal Society and Sir Gordon Cox, K.B.E.) and 6.17 (Professor S. H. Bauer, Cornell University).

Generally-accepted values for bond lengths, bond enthalpies, and spectroscopic absorptions have been used; they have been obtained from a variety of sources including the data books and spectroscopic text-books cited in this text. Accurate relative atomic masses are from "Mass and Abundance Tables for use in Mass Spectrometry", by J. H. Beynon and A. E. Williams, Elsevier, Amsterdam, 1963, and fragmentation patterns from "Compilation of Mass Spectral Data", by A. Cornu and R. Massot, Heyden, London, 1966.

I thank my colleagues, in particular, Dr. C. B. Thomas, Dr. D. J. Waddington and Mr. G. R. Walker (now Deputy Headmaster, Carisbrooke High School, Isle of Wight), for helpful discussions; I am very grateful to Dr. G. F. Liptrot (Head of Chemistry, Eton College) and Mr. H. S. Pickering (Uppingham School) for reading the manuscript and for making many helpful comments.

B. C. Gilbert

York 1975

Introduction

This text is written for A level students and for those taking introductory courses in Chemistry at degree and National Certificate levels.

The inclusion of some aspects of spectroscopic and diffraction techniques in A level syllabuses and early in degree and National Certificate courses is a trend which is surely to be welcomed. There have been significant advances in recent years in the use of these methods, not only for obtaining an analysis of the elements and groups present in a given molecule but also for establishing the way in which the constituent atoms are arranged. These developments have had a profound effect in increasing our scientific knowledge, especially in the fields of chemistry and molecular biology; this is reflected in the elucidation of the structure and mode of action of a wide range of compounds including, for example, antibiotics, enzymes, and nucleic acids.

It is important that such work should find its place in A level curricula—not only as a means of training potential chemists but also as part of an education to show, through the study of modern chemistry, how advances are made.

The subjects of spectroscopy and diffraction are often taught as elegant mathematical descriptions, but then the reasons for *using* the techniques are lost. In this text I have reduced the mathematical content to a minimum. In introducing mass spectrometry, infra-red, visible and ultra-violet spectroscopy, nuclear magnetic resonance, and diffraction, I have outlined in each chapter the theoretical essentials which highlight the most important principles, such as the quantum theory, the electromagnetic spectrum, and magnetism. However, it is the *results* from these techniques which are equally important at this stage. These are introduced very early in each chapter, so that it is possible from these data to see, at first hand, how the identity of a compound may be found and how its detailed molecular structure can be determined. The examples of the great majority of the compounds I have used are from A level syllabuses although I have given other examples of general interest to show how the techniques are being developed.

When studying these techniques, one so often forgets that they are but one set of methods which chemists use, and it is important for students to realize that although spectroscopic and diffraction methods provide very important evidence, it must be backed up by other data— in particular, by investigation of the chemical reactions of the compounds concerned. However, in a short text such as this it is not possible to remind ourselves of this on every occasion a technique is discussed, and I hope that, in reading this book, it will be realized how much we depend on other knowledge to provide us with definitive evidence for our conclusions. Such reminders are given in worked examples in the text and in the problems at the end of chapters.

Further reading is encouraged and references to articles and books, helpful for an A level or first year university student, are given. Some more advanced points are described but these are reproduced in small print so they can be omitted on the first reading if necessary.

S.I. Units have been used throughout and the nomenclature adopted is essentially the I.U.P.A.C. system as recommended by the Association for Science Education in their booklet "Chemical Nomenclature, Symbols and Terminology" (1972). An exception is the retention of the generic name ether [e.g. diethyl ether rather than ethoxyethane], as allowed by I.U.P.A.C. References to well known 'trivial' names accompany the recommended names (in parenthesis).

Chapter 1

Mass Spectrometry

The technique of Mass Spectrometry, which owes its origin to pioneering experiments carried out at the beginning of the present century, is now established as a means for obtaining the formulae and structures of molecules. It enables many sophisticated structural problems to be solved rapidly, even when only minute quantities of material are available.

1.1 THE MASS SPECTROMETRY EXPERIMENT

The principle of the method is to obtain a positively charged ion characteristic of the substance under investigation, and then, effectively, to determine the *mass* of this ion using an approach closely related to that employed by J. J. Thomson for measuring the charge-to-mass ratio (e/m) for cathode rays (electrons). The procedure involves the use of electric and magnetic fields to deflect charged particles.

In 1912, Thomson had demonstrated the use of a magnetic field to deflect a beam of positive ions, obtained by the ionisation (loss of an electron) of neon atoms. Close examination of the trace produced by the positive ions as they impinged on a detector demonstrated that there were two different types of ion, which were characterised as those from the two neon **isotopes** (^{20}Ne and ^{22}Ne). These differ in their masses because of the different numbers of neutrons in their nuclei. This experiment not only demonstrated the existence of isotopes but also laid the foundation for the development of mass spectrometry.

The design of a simple mass spectrometer is shown in Figure 1.1. The substance to be studied (a gas or the vapour from a relatively volatile liquid) is introduced into the ionisation chamber which is kept at very low pressure (about $10^{-4} N m^{-2}$). The vapour is bombarded with high energy electrons, and the collision between an electron and the molecule (or atom) under consideration causes an electron to be ejected from the latter, leaving a positively charged ion:

$$M \rightarrow M^+ + e^-$$

The positively charged ions are attracted by an applied electrostatic potential and are hence accelerated towards the negative plate. Ions are allowed to pass through a slit in the plate and the resultant beam is passed into a magnetic field; the positive ions then become deflected by an amount which depends upon the *mass* and the *charge* of each. The lighter the ion and the greater its charge, the greater will be the deflection.* The derivation of the exact relationship is as follows:

For an accelerating potential V, the *potential energy* of an ion, of charge e, generated in the ion-chamber is eV.

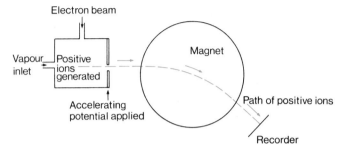

1.1   Basic features of a simple mass spectrometer

The ion is accelerated through the slit, and in this process its potential energy is completely converted into *kinetic energy*, $\frac{1}{2}mv^2$, where m and v are the mass and velocity, respectively, of the ion. Thus:

$$\tfrac{1}{2}mv^2 = eV \qquad (1.1)$$

When the positive ion passes into the magnetic field (of magnetic flux density B) it experiences a force at right angles both to the direction of motion and to the field direction. The magnitude of this force is Bev. The positive ion is now constrained to move in the arc of a circle of radius R, so that:

$$Bev = \frac{mv^2}{R} \qquad (1.2)$$

Combination of equations (1.1) and (1.2) leads to the following expression:

$$\frac{m}{e} = \frac{B^2R^2}{2V} \qquad (1.3)$$

This equation shows that for an ion of given mass (m) and charge (e) the radius of the circle of motion (R) is determined by B and V, i.e. the magnitudes of the magnetic and electric fields. In practice, R remains

*A simply-constructed model is described by A. Farmer, *The School Science Review*, 1969, **50**, 594.

fixed by the geometry of the apparatus and the position of the detector. It can then be seen that if V is kept constant and B is varied, equation (1.3) will be satisfied for ions of different m/e for different values of B. In other words, the value of B needed to get a particular type of ion to be deflected to the recorder is a measure of m/e for that ion.

Most positive ions generated will have lost just one electron (that is, M is ionised to M^+, rather than M^{2+} or M^{3+}), and they will there-fore have the same charge (opposite in sign, but equal in magnitude, to that of the electron). This means that, as B is varied, ions of different *mass* arrive at the recorder, and a spectrum of the masses of the various ions concerned can be plotted. Since the mass of the electron is very small compared with the mass of the nucleus (it is approximately one two-thousandth the mass of the proton) the experiment is effectively determining the masses of the parent molecules.

Early mass spectrometers worked almost exactly on this basis and achieved a reasonable resolution (separation of two fairly similar masses, such as ^{20}Ne and ^{22}Ne). Some modern mass spectrometers are designed with an extra focusing system which enables increased resolution to be obtained. As before, a narrow beam of positive ions with similar kinetic energies is produced. These ions have a small but finite range of kinetic energies, and this spread of energies must be reduced for more precise work so that equation (1.3) is strictly applic-able. This criterion is achieved by passing the beam of ions through an electric field (this is called an electrostatic analyser) which deflects the ions according to their kinetic energies (Figure 1.2). Then only one

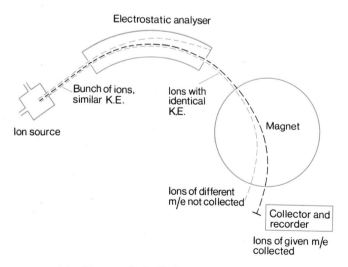

Electrostatic analyser

Bunch of ions, similar K.E.

Ions with identical K.E.

Ion source

Magnet

Ions of different m/e not collected

Collector and recorder

Ions of given m/e collected

1.2   Diagram of a double-focusing mass spectrometer

small component of the resulting beam, with a well-defined kinetic energy, is passed into the magnetic field for focusing of ions of given m/e values. As before, a scan of mass is obtained by varying B (although in certain circumstances it is possible to obtain a more rapid scan by varying V, with B constant). The detector usually comprises an electron multiplier as collector, and the arrival of ions appears as an electrical signal for display as a peak on an oscilloscope or recorder. This produces a **mass spectrum**—effectively a plot of the masses of the positive particles present against the relative number of ions of each mass. The scan is calibrated with a peak obtained from a substance of known relative atomic or molecular mass. The spectrum can usually be recorded in a few minutes, from less than a microgram of material.

Figure 1.3 shows the A.E.I. MS 50 Double-focusing Mass Spectrometer which operates on this basis and which can separate two masses differing by less than one part in 10^5.

1.2 MEASURING RELATIVE MOLECULAR AND ATOMIC MASSES

Modern mass spectrometers with a double-focusing facility can be used in two kinds of study. They may be employed to give fairly rapid scans of the relative masses of the ions from a variety of substances (this is also how a single-focusing instrument is employed). Alternatively, under carefully controlled conditions of high resolution, they can be used to separate closely spaced peaks and to determine the appropriate relative atomic and molecular masses with precision.

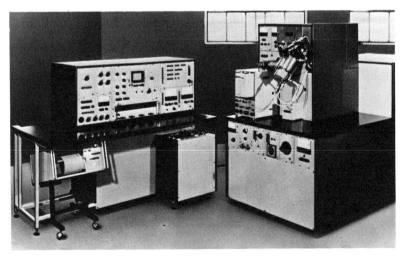

1.3 MS 50 Double-focusing mass spectrometer, A.E.I. Ltd. (Photo: Switchgear Testing Company, Manchester)

The following example should illustrate the advantages of the latter approach if the maximum amount of information is to be derived from an investigation. A peak which corresponds to mass 28 might be due to nitrogen ($^{14}N_2$), carbon monoxide ($^{12}C^{16}O$) or ethene ($^{12}C_2{}^1H_4$). However, these three molecules have slightly different relative molecular masses (these are based on the internationally accepted scale, with 12 exactly for the ^{12}C isotope):

$^{14}N_2$	28.0062
$^{12}C^{16}O$	27.9949
$^{12}C_2{}^1H_4$	28.0313

A modern mass spectrometer can readily be used to identify a peak exactly enough for it to be characterised as the positive ion from one of these. Further, if all three substances were to be present together, then under high-resolution conditions three separate peaks could be resolved. If one relative molecular mass is accurately known, this can be used to calibrate the field scan so that the other molecular masses can also be accurately determined.

Relative heights of separate peaks can also be used to obtain quantitative information. For example, from the mass spectrum of the monatomic gas neon can be measured not only the relative atomic masses of the constituent isotopes (^{20}Ne, ^{21}Ne, ^{22}Ne), to an accuracy of 1 part in 10^6, but also the relative abundance of the separate isotopes in the mixture:

	Relative atomic mass	Relative abundance %
^{20}Ne	19.9924	90.92
^{21}Ne	20.9940	0.26
^{22}Ne	21.9914	8.82

We must distinguish the *separate* isotopic atomic masses measured with the mass spectrometer from the *weighted average* obtained by other (chemical) methods. For example, ^{35}Cl has a relative atomic mass of 34.9688, and that of ^{37}Cl is 36.9659; the average atomic mass of the natural mixture of isotopes (75.53% ^{35}Cl, 24.47% ^{37}Cl) is 35.45).

These investigations lead on to very interesting questions about the causes of the widespread natural occurrence of certain isotopes of some elements (^{12}C, ^{16}O, for example) and also about the magnitude of the mass difference between a given isotope and the sum of its constituent neutrons and protons (this is closely related to the binding energy of the nucleus). However, our concern here will be with the rather different application of mass spectrometry to the determination of molecular structure.

1.3   MASS SPECTROMETRY OF MOLECULES

When an organic compound is introduced into the spectrometer, the molecules become ionised, by the loss of an electron, and the positive ions produced pass through the focusing system, leading usually to a peak at the appropriate relative molecular mass. However, the mass spectrum of an organic compound also contains extra information which can be extremely useful.

Figure 1.4 shows the mass spectrum of ethanol (CH_3CH_2OH); this is a plot of recorder signal (proportional to the number of ions of given m/e) against increasing m/e (in practice nearly all the ions have the same unit charge so this axis effectively corresponds to increasing mass; see page 11). The various traces indicated, (a)–(c), are recorded simultaneously with different degrees of amplification to allow the study of major and minor peaks.

Peaks occur at (or close to) most integral values (the extra precision possible with high resolution is not usually employed at this stage).

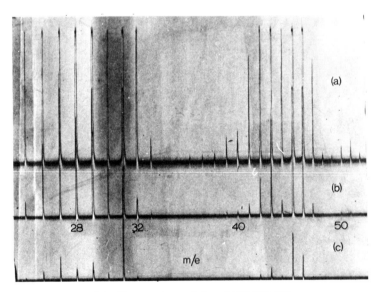

1.4   *Mass spectrum of ethanol, CH_3CH_2OH*

Many of the peaks are derived from ethanol by processes which will shortly be described. There are also peaks due to traces of air in the instrument; this gives rise to signals from N_2 (*m/e* 28) and O_2 (*m/e* 32), approximately in the expected ratio 4:1. These peaks may be used to calibrate the scan. For convenience, the spectrum is usually redrawn in somewhat simplified form (Figure 1.5) where only the major peaks from the organic compound are included. The peak heights are expressed as percentages of the height of the highest peak (the **base peak**), which is in this example the peak with *m/e* 31.

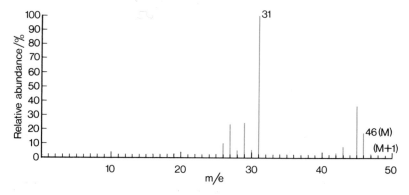

1.5   Stick diagram of the mass spectrum of ethanol

The mass spectrum of ethanol

The spectrum shows the expected peak at *m/e* 46, corresponding to the **molecular ion** (M^+) of the parent molecule (relative molecular mass is the sum of the relative atomic masses of $2C + 6H + O$).

There is also a very small peak at *m/e* 47, called the (M + 1) peak, which corresponds to the relatively few ethanol molecules present which, because they contain a ^{13}C, ^{17}O, or ^2H atom, have a molecular mass of 47. (^{13}C has a natural abundance, compared with ^{12}C, of 1.1%; for ^{17}O, relative to ^{16}O, the figure is 0.04% and for ^2H, relative to ^1H, the abundance is 0.01%.)* An even smaller (M + 2) peak arises from the molecules which contain two ^{13}C atoms, or an ^{18}O atom, or a ^{13}C and a ^{17}O atom, etc. It must be stressed that these all have the same chemical composition as ethanol, but they contain naturally occurring low-abundance isotopes.

---

* Isotope abundance tables can be found, for example, in "Book of Data: Chemistry, Physical Science, Physics", *Nuffield Advanced Science*, Penguin, London, 1972 (this also contains the accurate relative atomic masses of selected nuclei), and in "Chemistry Data Book", J. G. Stark and H. G. Wallace, SI Edn., John Murray, London, 1970.

The other peaks in the spectrum occur because some of the ethanol molecules which are first ionised to give M^+ then *fragment* to give smaller positively charged ions, a process which is understandable in terms of the high energy of the bombarding electrons. The positively charged fragments are accelerated in the usual way and are focused to be collected and recorded for their particular values of m/e.

The large peak in the mass spectrum of ethanol (the base peak) is at m/e 31; this corresponds to the positively charged fragment $[CH_2OH]^+$ obtained by loss of CH_3 from the parent ion $[CH_3CH_2OH]^{\ddagger}$. The structure of these and other positive ions will be discussed later (page 20), as will some guide-lines for interpreting fragmentation patterns, but it should at this stage be apparent that these peaks contain important clues about the structure of the molecule under investigation.

1.4 ANALYSIS OF MASS SPECTRA

The information of interest is contained mainly in the appearance of the peak from the molecular ion [with its associated (M + 1) and (M + 2) peaks] and in the fragmentation pattern. These features will now be considered in more detail.

(a) The Molecular Ion

For many molecules a peak of appreciable size from the molecular ion can be detected. It is usually a reliable guide that a molecule with π- or lone-pair electrons (e.g. benzene or ethanol, respectively) will give a detectable molecular ion (M) since one of these electrons can normally be lost (to give M^+) without the breakdown of the bonding framework in the molecule. However, because in some cases no peak from a molecular ion can be observed, care must be taken before it can be assumed that the peak at highest m/e is from the molecular ion.

It may be possible at this stage to determine very accurately the relative molecular mass of any given peak (if the high resolution facility is available) and this will then be carried out for the molecular ion itself. Because different atoms do not have exactly integral atomic masses and, in addition, because various combinations of similar mass are not identical (contrast, for example, C_2H_4 and N_2) the exact relative molecular mass (to 3 or 4 places of decimals) characterises the molecular formula exactly. For instance, a molecular peak with m/e almost exactly 60 could be from various possible molecules with different formulae, including $C_2H_4O_2$ [e.g. ethanoic acid (acetic acid), CH_3CO_2H] and C_3H_8O (e.g. propanol). The relative molecular masses of the molecular ions of these compounds are as follows:

$$C_2H_4O_2 \qquad\qquad 60.0211$$
$$C_3H_8O \qquad\qquad 60.0575$$

Under conditions of high resolution, these possibilities can be clearly distinguished. Further, we can extend this argument and show that given a precisely determined relative molecular mass we can obtain the molecular formula; for example, a molecular ion with m/e 94.0419 can be reliably attributed to a compound with the molecular formula C_6H_6O.

Information can often be extracted from the M, (M + 1), and (M + 2) peaks even if the high-resolution facility is not available or when demands are such that several straightforward scans are preferred to one very detailed examination. For example, for ethanoic acid ($C_2H_4O_2$), the height of the (M + 1) peak at m/e 61, which is mainly due to $^{12}C^{13}CH_4O_2$, should be just over 2% of the height of the peak from the molecular ion (m/e 60). This is because there is approximately a 2.2% chance that a molecule will contain one ^{13}C atom (there will be a much smaller percentage of molecules containing ^2H or ^{17}O). The corresponding figure for C_3H_8O is just over 3%, and for a compound with, say, eleven carbon atoms the relative intensities of M : (M + 1) peaks should be about 100 : 12. Clearly, then, measurement of the relative heights of the M and (M + 1) peaks, and sometimes of the (M + 2) peak, can be diagnostically useful, and extensive tables of accurate M : (M + 1) : (M + 2) ratios for various molecular formulae are available.* Sensible deductions can usually be made even if these details are not available. Thus the (M + 1) peak will be approximately N% of the main peak if the formula has N carbon atoms, a larger-than-usual (M + 2) peak may indicate that a sulphur atom is present in the molecule (^{34}S has a natural abundance of 4.22%), and so on.

In certain molecules the effect can be particularly striking, as illustrated by compounds containing chlorine or bromine atoms. Figure 1.6 is the mass spectrum of chloromethane; the peaks at m/e 50 and 52 are the molecular ions of $CH_3{}^{35}Cl$ and $CH_3{}^{37}Cl$, respectively, their relative intensities being in the ratio expected from the relative isotopic abundances of ^{35}Cl and ^{37}Cl (approximately 3 : 1). For

---

* An example of a portion of a Table giving accurate mass (M) and abundance [(M + 1), (M + 2)] figures for various formulae appears in "Physical Science: Students' Workbook II", *Nuffield Advanced Science*, Penguin, London 1972, p. 58.

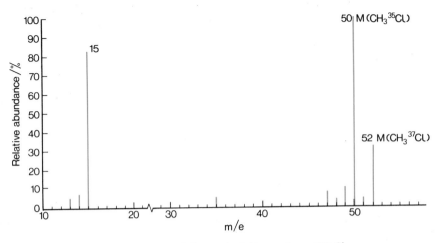

1.6   *Mass spectrum (stick diagram) of chloromethane, CH_3Cl*

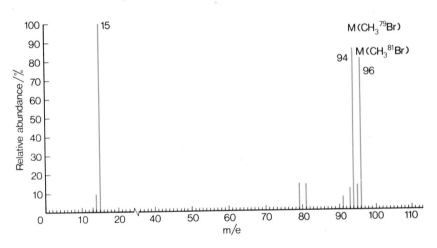

1.7   *Mass spectrum of bromomethane, CH_3Br*

bromomethane (Figure 1.7) the two almost equally intense peaks are from $CH_3{}^{79}Br$ and $CH_3{}^{81}Br$, the two bromine isotopes having approximately the same abundance.

Another useful rule is that a molecular ion with an *odd* value of m/e generally characterises a molecule with an odd number of nitrogen atoms.

It must be remembered that a peak of reasonable intensity at the highest m/e value observed is not necessarily from the molecular ion, but may be instead from the fragmentation pattern of a compound whose molecular ion has a peak too small to be clearly established.

Wherever possible, therefore, data should be interpreted together with information from other spectroscopic techniques and from conventional molecular mass and empirical and molecular formulae determination.

(b) Fragmentation Patterns

As already indicated, some of the molecules in the ionisation chamber fragment under the electron bombardment:

$$M \xrightarrow{-e} M^+ \longrightarrow P^+ + Q$$

A variety of fragmentation pathways is normally possible and for each route one of the fragments retains the positive charge. For example, another possibility here is fragmentation to $P + Q^+$, and further fragmentation of either P^+ or Q^+ may also occur.

The recognition of preferred modes of fragmentation (e.g. whether for this molecule a larger peak for P^+ or Q^+ is obtained) is assisted by practice with spectra from known molecules, but deductions are mainly based on chemical intuition, and a few simple guide-lines can be suggested. Most of the principles involved in recognising and predicting fragmentation patterns are closely related to those employed for discussing the chemistry of reactions in solution. For instance, we will need to consider which of a variety of possible fragments is best able to bear a positive charge, which bond is the weakest and therefore most likely to break, and which stable entities might readily be formed in simple decomposition pathways.

First, we must remember to do some elementary book-keeping with electrons and charges; most molecules have an even number of (paired) electrons, so that the positive molecular ion must have not only a *charge* but also an odd number of electrons. The fate of both the charge and the unpaired electron should be considered when fragmentation patterns are being interpreted.

The main types of fragmentation possible for a molecule may be characterised as follows.

(i) *Simple cleavage.* This involves the breakage of a single bond in the molecular ion, and a good example is provided by the mass spectrum of ethanol (Figures 1.4 and 1.5). The molecular ion is at m/e 46 and the base peak is at m/e 31; the latter corresponds to a molecular ion which has lost a group of mass 15 before being accelerated and focused. It is described as an $(M-15)$ peak, and is due to the ion $[CH_2OH]^+$ formed as follows:

$$[CH_3CH_2OH]^+ \longrightarrow \cdot CH_3 + [CH_2OH]^+$$
$$m/e \; 31$$

Note that the 'dot' indicates an unpaired electron. The products of this fragmentation then are the charged *ion* $[CH_2OH]^+$ (since the charge resides effectively on carbon this type of species is sometimes called a **carbonium** ion and written $^+CH_2OH$) and the neutral methyl **radical** $\cdot CH_3$. Only the former, being charged, is recorded in the mass spectrum.

Some of the other peaks arise as follows:

$$[C_2H_5OH]^+ \rightarrow H\cdot + [C_2H_5O]^+$$
$$m/e\ 45$$

$$[C_2H_5OH]^+ \rightarrow \cdot OH + [C_2H_5]^+$$
$$m/e\ 29$$

$$[C_2H_5OH]^+ \rightarrow H_2O + [C_2H_4]^+$$
$$m/e\ 28$$
$$\downarrow$$
$$H\cdot + [C_2H_3]^+$$
$$m/e\ 27$$

The peaks with m/e 27–29 are characteristic of a molecule containing an ethyl group, just as the appearance of an $(M - 15)$ peak is usually indicative of a methyl group in the parent compound. The reason that the peak with m/e 31, $[CH_2OH]^+$, is larger than those of the other positive ions is that this is a relatively *stable* positive ion compared to some of the other possibilities (e.g. $[CH_3]^+$, $[CH_3CH_2]^+$). This stability arises because oxygen has a lone-pair of electrons which can help to stabilise the positive charge on the adjacent carbon atom: this is possible because there is a spreading (delocalisation) of both the charge and the electrons between carbon and oxygen, a phenomenon which is indicated diagrammatically as follows:

$$^+CH_2 - \ddot{O} - H \quad \leftrightarrow \quad CH_2 \overset{+}{-} O - H$$

The use of the double-headed arrow implies that the actual electronic structure is somewhere in-between the two extremes indicated.

A similar explanation accounts for the formation of the fairly intense peak with m/e 45, attributed to the ion $^+CH(CH_3)OH$ formed by loss of a hydrogen atom (n.b. this ion is more stable than $CH_3CH_2O^+$).

This mode of fragmentation is often observed when an ion can be produced with the positive charge on a carbon atom adjacent to an atom with a lone-pair of electrons (e.g., O, S, N, halogen), and is often significant for alkanols, ethers, thiols, amines, and halogenoalkans. Their typical fragmentation patterns often allow these molecules to be recognised.

Other cases where stabilised positive ions are produced include

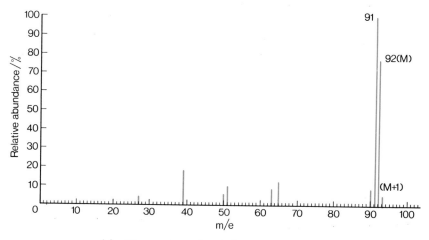

1.8   Mass spectrum of methylbenzene, $C_6H_5CH_3$

compounds containing the phenylmethyl (benzyl) group and also carbonyl-containing compounds. Figure 1.8 is the mass spectrum from methylbenzene (toluene) which illustrates the behaviour of compounds in the former group. (For this and the other spectra in the Chapter, only the ten most intense peaks are shown.) In addition to the molecular ion at m/e 92 and the (M + 1) peak at m/e 93 (which has an intensity which is nearly 8% of the molecular peak, since there are seven carbon atoms in methylbenzene; see page 15), there is an intense peak (the base peak) at m/e 91. This is from some of the methylbenzene molecules in the ionisation chamber which lose first an electron and then a hydrogen atom, to give the phenylmethyl cation as follows:

The reason for the comparative stability of this cation is the ease with which the *aromatic ring* can delocalise the positive charge, a phenomenon which can be represented diagrammatically as follows:

This type of fragmentation, to give a peak with $m/e = 91$, is characteristic of compounds of the type $C_6H_5CH_2X$. The mass spectrum from methylbenzene also shows other small peaks which indicate that alternative modes of fragmentation after high-energy bombardment are possible. These include the formation of two-, three- and four-carbon fragments (that with m/e 51 is $[C_4H_3]^+$), from fragmentation of the aromatic ring; the occurrence of these, though helpful, is not as diagnostically useful as the evidence from the main fragmentation pathway.

Carbonyl-containing compounds tend to decompose to give fragment ions where the positive charge is again shared between carbon and oxygen:

$$\left[\begin{array}{c} R \diagdown \diagup R' \\ C \\ \| \\ O \end{array}\right]^{\dot{+}} \longrightarrow R'\cdot + \left[R-\overset{+}{C}=O \longleftrightarrow R-C\equiv O^+\right]$$

Other peaks may arise because fragmentation can occur at the other C-alkyl bond (to give $R'-\overset{+}{C}=O$) and because these ions also tend to lose carbon monoxide quite readily:

$$[RCO]^+ \rightarrow [R]^+ + CO$$

The mass spectrum from butanone (methyl ethyl ketone) is shown in Figure 1.9. The important features to be noted in addition to the molecular ion (m/e 72) are the peaks at m/e 57 (M — 15) and 43 (M — 29), diagnostic of loss of $\cdot CH_3$ and $\cdot CH_2CH_3$ from the parent positive ion.

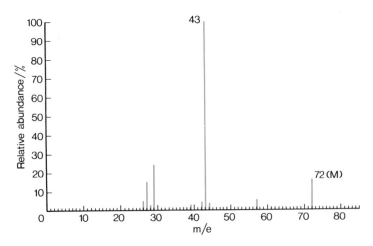

1.9   Mass spectrum of butanone, $CH_3COCH_2CH_3$

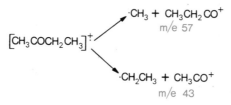

A peak at m/e 29 ($^+CH_2CH_3$) is also observed. The peaks at 57 (M−15) and 29 (M−43) help characterise a methyl-substituted ketone (loss of $\cdot CH_3$ and $CH_3CO\cdot$, respectively).

The spectrum of dimethylbutanone (methyl t-butyl ketone), Figure 1.10, shows the (M − 43) peak quite clearly. In this example the dimethylethyl (t-butyl) carbonium ion $^+C(CH_3)_3$, m/e 57, gives a particularly intense peak. This is because a **tertiary** carbonium ion (i.e. one with three alkyl groups attached to the carbon bearing the positive charge) is formed. This type of fragmentation tends to be favoured when such an ion can be obtained because a tertiary carbonium ion is more stable than a **secondary** carbonium ion, which is more stable than a **primary** carbonium ion (because of the overall electron-releasing property of alkyl groups). This means that fragmentation is generally preferred at the point of branching.

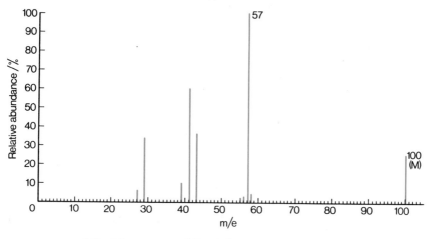

1.10   Mass spectrum of dimethylbutanone, $(CH_3)_3CCOCH_3$

(ii) *Fragmentation with rearrangement.* Occasionally a fragmentation process is detected which is rather more complicated than those discussed in section (i) above because molecular rearrangements are involved.

An example is provided by the mass spectrum of methyl butanoate ($CH_3CH_2CH_2CO_2CH_3$), which is shown in Figure 1.11. There is a trace of the expected molecular peak at m/e 102, and cleavage of the bonds to the carbonyl group leads to the peaks at m/e 43 (loss of $\cdot CO_2Me$), 71 (loss of $\cdot OCH_3$) and 59 (loss of $\cdot CH_2CH_2CH_3$):

$$
\left[CH_3CH_2CH_2C \begin{matrix} O \\ \backslash OCH_3 \end{matrix} \right]^{\dot{+}}
$$

$\nearrow \quad {}^+CH_2CH_2CH_3 + \cdot CO_2CH_3$
$\qquad \qquad m/e\ 43$

$\longrightarrow CH_3CH_2CH_2CO^+ + \cdot OCH_3$
$\qquad \qquad m/e\ 71$

$\searrow \quad \cdot CH_2CH_2CH_3 + {}^+CO_2CH_3$
$\qquad \qquad m/e\ 59$

1.11   Mass spectrum of methyl butanoate, $CH_3CH_2CH_2CO_2CH_3$

Other fragmentations lead to $CH_3CH_2CH_2CO_2{}^+$ (m/e 87) and to ${}^+CH_2CH_3$ (m/e 29). However, the unusual peak is that at m/e 74 (M − 28) which is thought to arise by transfer of a hydrogen atom to the suitably placed oxygen at the same time as fragmentation [the product ion (m/e 74) is recognisable as the enol* form of methyl ethanoate, $CH_3CO_2CH_3$]:

---

*Enol forms are discussed on page 81.

This is known as a **McLafferty** rearrangement and tends to occur when a hydrogen atom and a carbonyl oxygen come into close proximity. Inspection of models reveals that this can be achieved with the minimum of strain when there are six atoms in the chain.

(c)   Some examples of fragmentation patterns

The simple rules laid down so far should provide assistance with the solving of a variety of mass spectra. The following examples to a considerable extent typify the class of organic compound to which they belong.

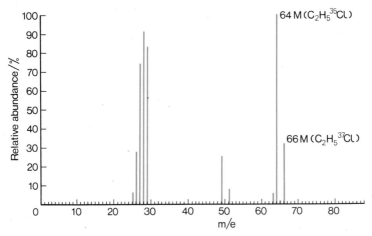

1.12   Mass spectrum of chloroethane, CH_3CH_2Cl

(i) *Chloroethane,* C_2H_5Cl (Figure 1.12). The mass spectrum shows the two expected molecular ions from $C_2H_5{}^{35}Cl$ and $C_2H_5{}^{37}Cl$, at m/e 64 and 66, respectively. The other main peaks are formed as follows:

$$[CH_3CH_2Cl]^+ \begin{cases} \longrightarrow \cdot H + {}^+CHClCH_3 \\ \qquad\qquad\qquad m/e\ 63,65 \\ \longrightarrow \cdot CH_3 + {}^+CH_2Cl \\ \qquad\qquad\qquad m/e\ 49,51 \\ \longrightarrow \cdot Cl + {}^+CH_2CH_3 \\ \qquad\qquad\qquad m/e\ 29 \\ \longrightarrow HCl + [C_2H_4]^+ \\ \qquad\qquad\qquad m/e\ 28 \end{cases}$$

The loss of H and CH_3 in the first two paths leaves the positive charge next to the chlorine (cf $^+CH_2OH$, $^+CH(CH_3)OH$, page 20). Another mode of fragmentation in this example involves elimination of the neutral molecule hydrogen chloride, to give a peak at m/e 28 from the positive ion from ethene (ethylene).

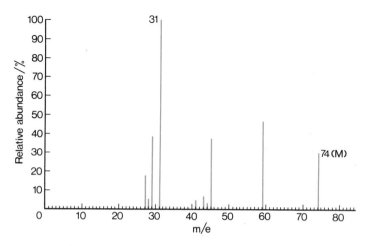

1.13   Mass spectrum of diethyl ether, $CH_3CH_2OCH_2CH_3$

(ii) *Diethyl ether (ethoxyethane)*, $CH_3CH_2OCH_2CH_3$ (Figure 1.13). The spectrum shows a molecular ion at m/e 74, and an $(M - 15)$ peak at m/e 59 which indicates loss of a methyl group; this leaves the fragment $^+CH_2OCH_2CH_3$ in which the oxygen atom is again able to exert a stabilising influence. The peaks at m/e 45 $(M - 29)$ and 29 are consistent with the occurrence of an ethyl group in the molecule; loss of this group gives either $^+CH_2CH_3$ and $\cdot OC_2H_5$ or $[C_2H_5O]^+$ and $\cdot CH_2CH_3$. The peak at m/e 31 is probably from $^+CH_2OH$, arising as follows:

Similarly,

$$^+CH_2-O \overset{}{\lessgtr} CH_2$$
$$\underset{H \overset{}{\lessgtr} CH_2}{|} \longrightarrow \ ^+CH_2OH + CH_2{=}CH_2$$
$$m/e \ 31$$

$$CH_3\overset{+}{C}H-O-CH_2CH_3 \longrightarrow CH_3\overset{+}{C}HOH + CH_2{=}CH_2$$
$$m/e \ 45$$

The driving force for these fragmentations is the production of a stable molecule (ethene), with the retention of the positive charge next to oxygen in the remaining cation.

(iii) *Diethylamine,* $(C_2H_5)_2NH$ (Figure 1.14). Particularly important here is the odd-numbered molecular ion $(m/e$ 73), which confirms a structure with a single nitrogen atom. Fragmentation in this example leads to peaks from $^+CH(CH_3)NHCH_2CH_3$ $(m/e$ 72) and

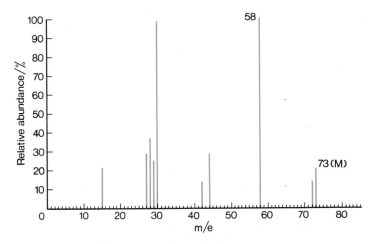

1.14   Mass spectrum of diethylamine, $(CH_3CH_2)_2NH$

$^+CH_2NHCH_2CH_3$ [m/e 58 (M − 15)] both of which have the positive charge adjacent to the nitrogen atom with its lone pair of electrons. Loss of an ethyl group is indicated by the (M − 29) peak at m/e 44; this peak could be from $^+NHCH_2CH_3$, but the rearranged (stabilised) isomer $^+CH(CH_3)NH_2$ seems more likely. The peak at m/e 30 is probably from $^+CH_2NH_2$, formed by a complex fragmentation-rearrangement process (*cf.* $^+CH_2OH$ in the preceding example).

(iv) *Phenylethanone (acetophenone)*, $C_6H_5COCH_3$ (Figure 1.15).

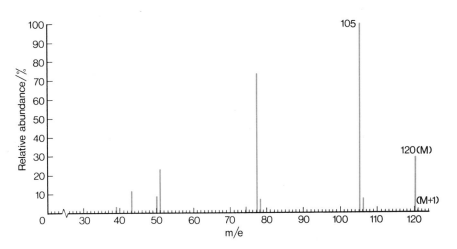

1.15   Mass spectrum of phenylethanone, $C_6H_5COCH_3$

This mass spectrum shows a molecular ion at m/e 120 and a characteristic (M − 15) peak (*m/e* 105) from loss of a methyl group. Fragmentation at the other side of the carbonyl group also occurs, to give the (M − 43) peak at *m/e* 77:

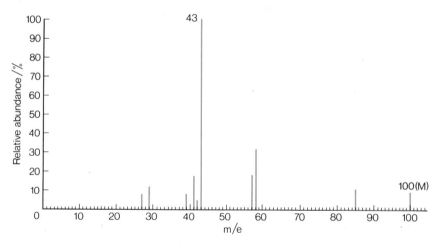

The detection of a peak at *m/e* 77 (and, in general, peaks in the 75–77 region) provides very strong evidence for a benzenoid compound. This is also true to a certain extent for the fairly prominent peaks at *m/e* 50 and 51 formed in the breakdown of the aromatic ring.

1.16   Mass spectrum of 4-methylpentan-2-one, CH₃COCH₂ CH(CH₃)₂

(v) *4-Methylpentan-2-one (methyl isobutyl ketone)*, (Figure 1.16) $CH_3COCH_2CH(CH_3)_2$. In this example the peaks at *m/e* 85 [(M − 15)], 43, and 57 [(M − 43)] confirm the presence of the COCH₃ group. The peak from the $^+CH_2CH(CH_3)_2$ fragment (*m/e* 57) is not so dominant as that of $^+C(CH_3)_3$ in the isomeric ketone previously considered (Figure 1.10), which is as expected for the lower stability of the primary $[^+CH_2CH(CH_3)_2]$ rather than the tertiary $[^+C(CH_3)_3]$ ion. The peak at m/e 58 (M − 42) arises via a McLafferty rearrangement, as follows:

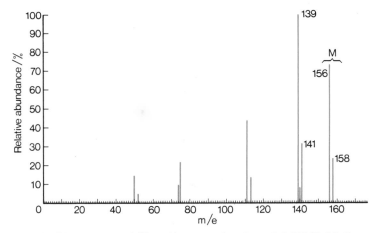

(vi) *Chloro-4-benzenecarboxylic acid (4-chlorobenzoic acid)*, 4-ClC₆H₄CO₂H (Figure 1.17). This spectrum shows features as expected for a compound with two functional groups. First, attention is drawn to the two peaks at high m/e values, 156 and 158, in the intensity ratio of approximately 3 : 1 and characteristic of a chlorine-containing compound. Then, the peak at m/e 75 confirms the aromatic

1.17 Mass spectrum of chloro-4-benzenecarboxylic acid, 4-ClC₆H₄CO₂H

nature of the compound. The peaks at m/e 139 and 141, (M − 17), again in the ratio 3 : 1, indicate that fragmentation has taken place (probably with loss of OH) with retention of chlorine in the positive ion. The chlorine is also retained for the fragments at m/e 111/113. The following scheme is the likely process:

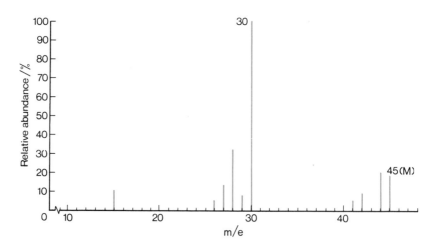

It should be explained here that the mass spectra of the 2- and 3-isomers of this compound would resemble somewhat that of the 4-isomer and therefore that mass spectrometry might not provide unambiguous assignment of a spectrum to one particular isomer. However, structure determination will usually be carried out with a variety of techniques and, for example, the nuclear magnetic resonance and infra-red spectra of the isomers, when examined in conjunction with the mass spectrometry evidence, would normally enable the distinction to be made.

1.5 WORKED EXAMPLES

(a) Spectra 1–3
Figures 1.18–1.20 are the mass spectra of three unknown compounds. In each case the molecular ion (or ions, for spectrum 1.19) and the base peak are denoted, and the ten most intense peaks have been included. With no further information provided, can you identify the compounds?

1.18   Mass spectrum of Worked Example 1.1

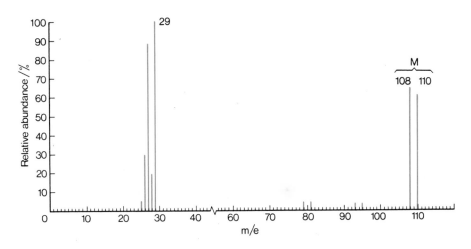

1.19   Mass spectrum of Worked Example 1.2

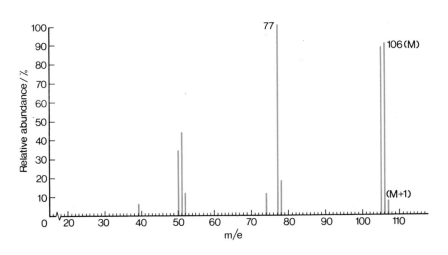

1.20   Mass spectrum of Worked Example 1.3

(b) **Discussion**

(i) *Example 1.1.* Figure 1.18 is the mass spectrum of ethylamine, $CH_3CH_2NH_2$. The important features to note are the odd-numbered molecular ion, indicating a molecule containing a single nitrogen atom, the intense $(M-1)$ and $(M-15)$ peaks, and the typical "ethyl-group" peaks at 26–29. All this supports a molecular formula C_2H_7N (rather than say CH_3NO). The breakdown pattern observed derives, in part, from the ability of nitrogen to stabilise an adjacent carbonium ion, and loss of a stable molecule (ammonia in this example) is also observed.

$$[CH_3CH_2NH_2]^+ \begin{array}{l} \nearrow H\cdot + {}^+CH(CH_3)NH_2 \\ \quad m/e\ 44 \\ \rightarrow \cdot CH_3 + {}^+CH_2NH_2 \\ \quad m/e\ 30 \\ \searrow NH_3 + [CH_2=CH_2]^+ \\ \quad m/e\ 28 \end{array}$$

(ii) *Example 1.2.* Figure 1.19 is the mass spectrum of bromoethane, CH_3CH_2Br. The crucial evidence here lies in the "double" molecular peak, two units of m/e apart, from molecules containing ^{79}Br and ^{81}Br (in almost equal abundance). The peak at m/e 29 strongly suggests an ethyl group, which is confirmed by the peaks from m/e 25 to 28. Other fragments which can be recognised are the bromine atoms themselves (m/e 79/81) and also the $(M-15)$ peaks (m/e 93/95).

$$[CH_3CH_2Br]^+ \begin{array}{l} \nearrow {}^+CH_2CH_3 + Br\cdot \\ \quad m/e\ 29 \\ \rightarrow \cdot CH_2CH_3 + Br^+ \\ \quad m/e\ 79,\ 81 \\ \searrow \cdot CH_3 + {}^+CH_2Br \\ \quad m/e\ 93,\ 95 \end{array}$$

Comparison of this spectrum with that from chloroethane (Figure 1.12) indicates a greater relative extent of halogen loss in this example, which is consistent with the C—Br bond being weaker than the C—Cl bond.

(iii) *Example 1.3.* As judged by the mass spectrum (Figure 1.20), this compound is aromatic (there is large peak at m/e 77 and also peaks at 50/51) and it can easily lose one hydrogen atom (to give the large peak at 105; the relative molecular mass is 106). Two possibilities which could lose a fragment of mass 29 to leave the peak at m/e 77 are ethylbenzene and benzaldehyde (benzenecarbaldehyde). However, the former would be expected to give an intense peak at 91 (from phenyl-methyl, $[C_6H_5\!-\!CH_2]^+$; see page 21) so the latter structure is preferred. The loss of a hydrogen atom is expected to be particularly facile for benzaldehyde:

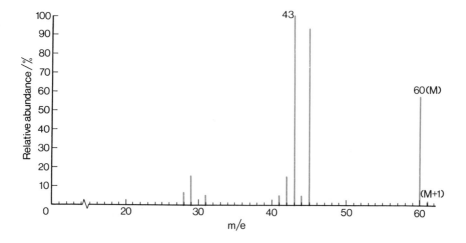

1.6 PROBLEMS

1.1 Identify the compound whose mass spectrum is shown in Figure 1.21. The peak at m/e 61 is approximately 2% as high as that at m/e 60.

1.21   Mass spectrum for Problem 1.1

1.2 Predict the relative intensities and m/e values of the molecular ions of different isotopic composition in the mass spectrum of 1, 1-dibromoethane, CH_3CHBr_2.

1.3 Identify the compound (of molecular formula $C_9H_{10}O_2$) whose mass spectrum is shown in Figure 1.22.

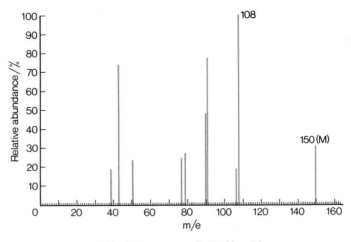

1.22   Mass spectrum for Problem 1.3

How do you account for the formation of a fragment with m/e 108?

(Infra-red and nuclear magnetic resonance spectra of this compound appear on pages 66 and 114 respectively).

More problems involving mass spectra appear at the end of the book.

1.7 DEVELOPMENTS AND APPLICATIONS

(a) Mass Spectrometry and Gas Chromatography

The previous sections demonstrate how the mass spectrometer provides an extremely sensitive technique for determining the molecular formulae and structures of a wide variety of organic compounds. A recent development provides, in addition, a facility for the rapid analysis of complicated *mixtures* of different compounds.

A small quantity of the mixture to be examined is dissolved in a volatile solvent (e.g. diethyl ether), and the solution is injected into a gas chromatograph. The chromatograph separates the various volatile components, which emerge from the chromatographic column in the carrier gas stream after different times have elapsed (depending on their retention times on the column). The effluent is then divided into two streams, one of which passes to a detector and a recorder which indicates with a peak when a compound is in the stream emerging from the column; a sample of this pure component is fed into the mass spectrometer from the second stream and its spectrum is recorded with a rapid scan. In this way the *chromatograph* achieves the separation of components (the areas under the chromatographic traces indicate the relative amounts of the separate components in the mixture) and the *mass spectrometer* provides the additional information which leads to the relative molecular masses and structures of the separate components. The combination of the two techniques provides an excellent method for determining the products, and their relative yields, from an organic reaction.

(b) Non-volatile compounds

One of the limitations of the mass spectrometer is that it is only suitable for those materials which are easily vaporised (i.e. which are volatile). Substances which may be unsuitable include salts and organic compounds of high relative molecular mass or of a polar nature, like many carboxylic (alkanoic) acids and compounds with several hydroxyl groups (e.g. carbohydrates). However, one way to overcome this problem is to prepare a suitable chemical derivative of higher volatility than the compound in question. This can be achieved by converting acids into esters ($RCOOH \rightarrow RCOOCH_3$), alkanols (ROH) into alkoxy derivatives (e.g. $ROCH_3$) or trimethylsilyl derivatives ($ROSiMe_3$), and amides into *N*-alkylated derivatives. For example, methylation of the N—H groups in some small peptides allows their mass spectra to be successfully investigated, because the intermolecular hydrogen bonding between the amido hydrogen atoms and carbonyl oxygen atoms (NH-----O$=$C), which normally renders the molecules non-volatile, is disrupted.

This approach considerably extends the range of problems which can be tackled with mass spectrometry and provides a rapid method for obtaining the amino-acid sequence in some small polypeptides **(peptide sequencing)**. Thus, a peptide can be represented diagrammatically as A—B—C—D— $\cdots$ where each letter corresponds to one of the possible amino acids, joined by the **peptide link** (—NH—CO—). Methylation, followed by investigation with mass spectrometry, leads to a variety of fragment ions of the type AB^+, ABC^+, $ABCD^+$, etc., each of which can be recognised because the masses of the individual (methylated) amino acids are known (this type of analysis can often be computer-aided). In this way the fragments and, ultimately, the amino-acid sequence, can be determined. This method can be employed for small quantities ($\sim$ mg) of peptides with up to ten amino-acid residues (these peptides could themselves be fragments of a larger polypeptide or protein). In terms of the time and the quantities of material involved, this technique clearly has advantages over a more traditional approach employing the sequential removal and chemical analysis of end-groups (terminal-group analysis).

(c) Isotope labelling

The ready detection of different isotopes of a given element with a mass spectrometer leads to its use in conjunction with **isotope labelling** for a range of applications, including the investigation of organic reaction mechanisms.

For example, suppose a research programme is initiated to diagnose the mechanism of hydrolysis of an ester under particular conditions. It can be envisaged that either of the following mechanisms [involving acyl-oxygen fission, (i), or alkyl-oxygen fission, (ii)] might be followed.

If the parent ester with an isotopic label (e.g. ^{18}O at the position indicated) is prepared, then, depending on the mechanism, either the alkanol [route (i)] or the acid [route (ii)] would show a molecular peak two units higher than that for the unlabelled compound. It is not necessary to go to the trouble and expense of using 100% enrichment at the particular position, since a much smaller extent of enrichment can be employed and "followed" through to the products. An alternative approach to this kind of information involves an appropriate radioactive isotope (^{14}C is widely employed) and a Geiger counter to identify which of the products contains the radioactive label.

(d) Other information

Attention can also be focused upon the processes which take place when electrons collide with molecules in the ionisation chamber. For example, by studying the appearance of various peaks in a mass spectrum as the energy of the bombarding electrons is increased it is possible to determine the ionisation energy of the molecule (i.e. the threshold energy at which the collision knocks out an electron) and also the dissociation enthalpy (energy) of a bond which is broken in a fragmentation process. This branch of mass spectrometry is called **electron impact.**

Other features to be observed in mass spectra are, occasionally, peaks from doubly charged ions (M^{2+}, with $e = 2$) and also broad **metastable** peaks which may also be recognised because they sometimes occur at other than the usual nearly-integral mass numbers. These peaks arise from ions which fragment in the region *between* the electrostatic analyser and the magnetic field, rather than in the ionisation chamber.

$$PQ^+ \longrightarrow P^+ + Q\cdot$$

parent	daughter
ion	ion

The simple relationships derived earlier (page 10) are not now obeyed; if the masses of the **parent** (PQ^+) and **daughter** (P^+) ions are *a* and *b*, respectively, then it can be shown that the broad peak which characterises the arrival of P^+ at the recorder has $m/e = b^2/a$. An example is a broad peak which is found at m/e 56.5 in the mass spectrum of phenylethanone (see page 27). This arises from the decomposition of ions of m/e 105 to give the positive-ion fragment with m/e 77 ($77^2/105 = 56.5$). In the same spectrum another metastable peak is observed with m/e 33.8; the reader might like to ascertain the relevant fragmentation process.

Further Reading

H. C. Hill, "Introduction to Mass Spectrometry", 2nd Edn. (revised by A. G. Loudon), Heydon and Sons, London, 1972.

R. I. Reed, "The Mass Spectrometer in Organic Chemistry", *Quarterly Reviews of the Chemical Society*, 1966, **20**, 527.

G. W. A. Milne, "The Application of High Resolution Mass Spectroscopy to Organic Chemistry", *Quarterly Reviews of the Chemical Society*, 1968, **22**, 75.

M. M. Campbell and O. Runquist, "Fragmentation Mechanisms in Mass Spectrometry", *J. Chemical Education*, 1972, **49**, 104.

J. F. J. Todd, "Modern Aspects of Mass Spectrometry", *Education in Chemistry*, 1973, **10**, 89.

More advanced reading (these books also provide descriptions of some of the techniques described in other chapters).

D. H. Williams and I. Fleming, "Spectroscopic Methods in Organic Chemistry", 2nd Edn., McGraw-Hill, London, 1973.

R. M. Silverstein, C. G. Bassler, and T. C. Morrill, "Spectrometric Identification of Organic Compounds", 3rd Edn., Wiley, New York, 1974.

S. F. Dyke, A. J. Floyd, M. Sainsbury, and R. S. Theobald, "Organic Spectroscopy: an Introduction", Penguin, London, 1971.

Film

"Analysis by Mass", A.E.I. Ltd. Available from Guild Sound and Vision Ltd., Peterborough.

Chapter 2

Introduction to Spectroscopic Techniques

Energy is absorbed or emitted by molecules and atoms in discrete amounts, corresponding to precise changes in energy of the molecule or atom concerned; this is a fundamental part of the **quantum theory**. It is possible both to verify that this behaviour occurs and to measure the amounts of energy involved; this is so because when a certain amount of energy is emitted (for example), the energy appears as electromagnetic radiation of a precise frequency, given by the following relationship

$$\Delta E = h\nu \tag{2.1}$$

where ΔE is the change of energy involved,
ν is the frequency of the radiation,
and h is Planck's constant $(6.626 \times 10^{-34} \, \text{J s})$.
As will be seen in the following chapters, molecules and atoms are found to absorb only certain frequencies but not others; similarly, under certain circumstances, discrete frequencies in different parts of the electromagnetic spectrum are emitted and can be identified.

2.1 ELECTROMAGNETIC RADIATION: ENERGY, FREQUENCY, AND WAVELENGTH

The realisation that visible light is just one of the possible forms of electromagnetic radiation is essential to our understanding of spectroscopic and diffraction methods for studying molecular structure. Other forms of this radiation—like X-rays, ultra-violet and infra-red radiation, microwaves and radiowaves—may be manifest in different ways and may have different applications, but they are essentially the same phenomenon. All forms are characterised by their **frequency** (and hence **wavelength**) and their **energy**.

That different types of radiation possess different energies is not difficult to appreciate (X-rays and ultra-violet radiation can cause much more damage to human tissues than does visible radiation) and it can also be demonstrated convincingly that radiation is wave-like in nature. For example, when light from a single source passes through two narrow parallel slits and then falls on a screen behind the slits, an interference pattern of dark and light lines is produced (Young's slits experiment). In an analogous fashion it can be shown that when a beam of X-rays is incident on a powdered solid, the atoms or ions in the latter behave as a diffraction grating and an interference pattern of maxima and minima is obtained in the reflected X-ray beam. The eye cannot see the interference pattern because X-rays are not visible, but a photographic film is darkened by the reflected beam and reveals the pattern resulting from the interference.

Electromagnetic radiation consists of oscillating electric and magnetic fields which can be transmitted through space. In a given medium all electromagnetic radiation travels with the same velocity (for example, 3×10^8 m s^{-1} in a vacuum). The oscillations associated with different types of electromagnetic radiation proceed in a wave-like fashion with different *wavelengths* (the distance between successive peaks) and *frequencies* (the number of waves passing a given point per second:

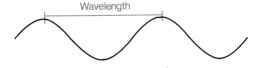

The relationship between c, the velocity, λ, the wavelength, and ν, the frequency is:

$$c = \lambda \times \nu \qquad (2.2)$$

λ has the units of length (normally m) and ν has the units of time^{-1} [normally s^{-1} or Hz (Hertz)].

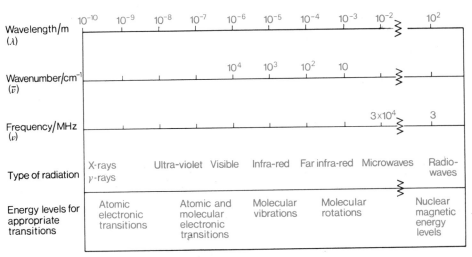

2.1 The electromagnetic spectrum

Figure 2.1 summarises the approximate ranges of the wavelengths for the different parts of the electromagnetic spectrum.* X-rays, for example, are characterised by a very short wavelength (λ is about 10^{-9} m, i.e. 1 nm) and a correspondingly high frequency and energy. Visible light has longer wavelengths, from about 400 nm (which is violet light) to about 750 nm (the red end of the visible part of the spectrum). Radiowaves have much greater wavelengths—of the order of metres.

In some parts of the spectrum the radiation is customarily referred to in terms of its frequency; radiowaves with $\lambda = 100$ m have $\nu = 3 \times 10^6$ Hz, i.e. 3 MHz.

Reference may also be made (especially in the infra-red region) to the reciprocal wavelength, known as the **wavenumber** ($\bar{\nu}$); like frequency, wavenumber is directly proportional to energy. Because of its almost universal usage for infra-red radiation the wavenumber will be retained here in that context. The unit customarily employed is cm^{-1}; if $\lambda = 10^{-6}$ m, the wavenumber, $\bar{\nu}$, is $10^4 cm^{-1}$.

2.2 ATOMIC AND MOLECULAR ENERGY LEVELS

(a) **Atoms**

Our understanding of the existence of well-defined energy levels for electrons in atoms originates in spectroscopic observations on the radiation emitted when atoms or ions become excited. The essential part of the experiment is illustrated by the flame test for alkali metals: for example, a small quantity of a sodium salt, when placed in a flame, appears yellow. This is because sodium atoms are formed and, at the high temperature of the flame, these become excited by absorption of energy: this energy is subsequently emitted as radiation in the visible region. The experiment can be carried out more precisely by analysing the emitted radiation in terms of the frequencies (and hence wavelengths) present: the radiation is incident on a prism which disperses radiation of different wavelengths in different directions (like Newton's experiment to produce all the colours in the visible spectrum from "white" light). The instrument which is used to investigate this is called a spectroscope (hence the term *spectroscopy* for studies of this kind) and the resulting radiation (not all of which will be in the visible region) is detected with a photographic film.

---

* Detailed Tables also appear in "Book of Data: Chemistry, Physical Sciences, Physics", *Nuffield Advanced Science*, Penguin, London, 1972, pp. 42–45, and in "Chemistry Data Book", J. G. Stark and H. G. Wallace, SI Edn., John Murray, London, 1970, p. 2.

It can then be demonstrated that from excited atoms only certain wavelengths are emitted (the resulting plot of wavelength contains several discrete values or "lines"—these are typical line-spectra*). In the example of the sodium flame (and, in terms of a useful application, the sodium-vapour street lamp), the orange colour derives mainly from the intense emission with $\lambda = 589$ nm.

Line-spectra can be used to identify different elements present (e.g. in a vaporised metal sample) and also, under certain circumstances, to determine the amount of a particular material present (from the intensity of the radiation associated with a particular transition).

The spectra can be interpreted in terms of the energy levels associated with electrons in various shells (or orbitals) in the atom concerned. For example, the emission with $\lambda = 589$ nm for the sodium atom corresponds to the energy emitted when an electron which has been promoted to a $3p$ orbital returns to the lower energy $3s$ orbital. The energy change can be larger if "inner" electrons are involved; for example, bombardment of copper with electrons leads to the ejection of an inner ($1s$) electron and the resulting transition of an outer electron from a higher energy level down to this vacancy, for example from $2p$ to $1s$, leads to the emission of X-rays ($\lambda = 0.15$ nm in this case).

Absorption of exact amounts of energy—a process leading to a transition *from* a lower *to* a higher energy level—is also possible. A clear example here is provided by the **Fraunhofer** lines: in the otherwise continuous spectrum of visible radiation from the sun, several wavelengths are missing. These correspond exactly to lines in the emission spectra of various atoms, including hydrogen. The explanation is that the continuous radiation from the sun passes through its atmosphere, which contains these atoms; the atoms absorb energy of discrete wavelengths which are then missing from the radiation reaching the earth. This provides a method for detecting gases in the atmosphere of the sun.

(b) Molecules

It can also be demonstrated experimentally that *molecules* have certain exact energy levels and that absorption of discrete frequencies (and hence energies) from incident radiation raises molecules from lower energy levels to upper energy levels. In general, there are three areas of the electromagnetic spectrum with which we will be concerned.

---

* Several examples appear on pp. 46 and 47 in the "Book of Data", referred to on p. 40 of this chapter, and also (in the appropriate colours) on the front inside cover of "Chemistry: Students' Book I", *Nuffield Advanced Science*, Penguin, London, 1970.

First, the electrons in molecules occupy molecular orbitals with precise energy levels (cf. electrons in atomic orbitals). Transitions from lower filled orbitals to upper (higher energy) empty orbitals usually involve absorption of radiation in the *ultra-violet* (u.v.) and *visible* parts of the spectrum. This is the basis of electronic absorption spectroscopy (Chapter 4).

Much smaller quantities of energy are associated with changes in the rotational energy of a molecule (which is allowed only certain well-defined values) and in its vibrational energy (which is also quantised). This second area of interest is concerned with precise energy absorption in the *infra-red* (i.r.) part of the spectrum. How the measurement and interpretation of these energy changes in molecules leads to structural information is described in detail in Chapter 3.

The energy changes associated with certain nuclei in magnetic fields (this is the basis of *nuclear magnetic resonance* spectroscopy, n.m.r) are smaller still and occur in the radio-frequency region. This type of spectroscopy (see Chapter 5) differs from the two forms mentioned above in that a magnetic field has to be applied before distinct energy levels for these nuclei are established.

It is the differences in the magnitudes of the energy changes involved which dictate the necessity for different instrumental arrangements for u.v. (and visible), i.r., and n.m.r. spectroscopy; these will be described in the following chapters.

Finally, it is the realisation of the origin of a particular wavelength of radiation emitted (or absorbed) which leads to detailed information about the molecules involved and provides a key to the investigation of molecular structure.

Introductory Reading
"Chemistry: Students' Book I", *Nuffield Advanced Science*, Penguin, London, 1970:
Topic 4 (Atomic Structure and Spectra).
Topic 8 (Waves, Interference, Electron- and X-ray Diffraction, Infra-red Spectra).

Basic Principles and Experimental Techniques
R. J. Taylor, "The Physics of Chemical Structure", Revised Edn., *Unilever Educational Booklet*, Unilever Education, London, 1967.

Further Reading
"Physical Science: Students' Workbook II", *Nuffield Advanced Science*, Penguin, London, 1972:
Section 14 (Electromagnetic radiation, Waves, Interference, Diffraction, Quantum Theory).
J. H. J. Peet, "A Survey of Molecular Spectroscopy", *The School Science Review*, 1972, **54**, 281.

Chapter 3

Infra-red Spectroscopy

Infra-red (i.r.) radiation is the term used to describe electromagnetic radiation with frequencies and energies somewhat lower than those associated with visible light (see page 39) and is emitted as a range of frequencies and energies from a hot body (sometimes together with visible radiation).

When a beam of i.r. radiation is incident upon a collection of certain molecules, absorption of discrete frequencies by the molecules takes place. This corresponds to the absorption of well-defined amounts of *energy* by the molecules from the spread of energies which constitutes the radiation, and there are two main types of application of i.r. absorption spectroscopy which provide important structural information about the molecules concerned.

The first is the study of simple molecules (usually diatomic and triatomic molecules) in the gas phase; it is possible to relate the exact amounts of energy absorbed from the i.r. radiation to increases in the rotational and vibrational energy of the molecules (Figures 3.1 and 3.7, on pages 44 and 53, respectively, illustrate the molecular motions involved). From this type of investigation it is possible to determine *bond lengths* and also *force constants* (which are a measure of the resistance of bonds to stretching).

An understanding of these features is a suitable basis for the application of i.r. spectroscopy in the second important field, namely the recognition of the *structures* of more complicated molecules from their characteristic absorptions. As with mass spectrometry and n.m.r. spectroscopy (Chapter 5), i.r. spectroscopy can be used to indicate the nature of the functional groups in a molecule, and, by comparison with spectra from known compounds, to provide assistance in the identification of an unknown compound.

Although the previous dominant position of i.r. spectroscopy for structure determination has been inherited by mass spectrometry and n.m.r. spectroscopy, it is nevertheless a very useful accessory. Compared with the other techniques, the instrumentation is inexpensive and operation is straightforward.

3.1 PURE ROTATION I.R. SPECTRA OF SMALL MOLECULES

For gaseous diatomic molecules, an absorption of i.r. radiation is only detected if the molecule has a **dipole moment**. This occurs when the two atoms are chemically different, such that an unequal sharing of electrons leads to an asymmetric distribution of electron density, which is denoted as follows:

$$\overset{\delta+}{H} - \overset{\delta-}{F}$$

The molecule is rotating, both about an axis along the bond and also about an axis through the centre of gravity and perpendicular to the bond (Figure 3.1). The latter motion gives rise to a fluctuating electric field which enables this type of molecule to interact with the fluctuating electric field of the incoming radiation and hence, by absorbing energy from the radiation, to increase its rotational energy.*

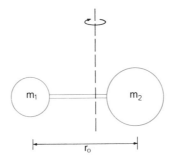

3.1   *Simple model for the rotational motion of a diatomic molecule*

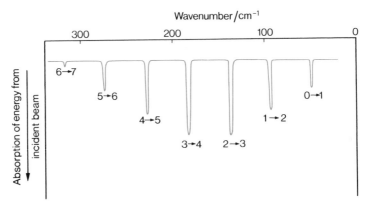

3.2   *Rotational i.r. absorption spectrum (in the range* $\bar{\nu}$ *0–300 cm^{-1}) for gaseous hydrogen fluoride, HF.*

---

* Molecules without dipoles rotate (and vibrate) in similar fashion, but do not interact in this way with incident radiation.

Figure 3.2 shows the spectrum obtained when radiation with wavenumber ($\bar{v}$) in the 0–300 cm^{-1} range is incident upon a gaseous sample of HF; there are several absorptions of energy, recognised by the downward peaks, at characteristic wavenumbers (i.e. at certain exact energies: E is proportional to $\bar{v}$). It is important to note that the separation between the lines is approximately constant (40.5 cm^{-1}).

Similar spectra, but with different separations, are obtained for other heteronuclear diatomic molecules (for HCl, for example, the spacing is 20.7 cm^{-1}).

The explanation for the appearance of the spectra is as follows. For a molecule which consists of two masses m_1 and m_2 and a bond length r_0, the only allowed values for the rotational energy (E_{rot}) are given by the expression

$$E_{rot} = \frac{h^2 J(J+1)}{8\pi^2 I} \tag{3.1}$$

where h is Planck's constant (6.626 × 10^{-34} Js) and where I and J require further explanation:

I is the *moment of inertia* for rotation about the axis indicated in Figure 3.1 and is given by $[m_1 m_2/(m_1 + m_2)]r_0^2$.

J is a *quantum number*, with the allowed values of 0, 1, 2, 3 . . . . Only certain rotational energy levels (E_{rot}) for the molecule are allowed, because the quantum number is restricted to certain values: the energy levels are indicated in Figure 3.3. Each molecule at any moment must be in one of these energy levels, and since the energy separations between the levels are small, all the states indicated will be fairly well populated (the problem of distribution of molecules amongst energy levels is discussed later).

A molecule can increase its rotational energy by being promoted from one level to the next level (i.e. $\Delta J = 1$) if *exactly* the correct amount of energy is incident upon it. Thus, those specific wavenumbers whose energies correspond to the absorptions of energy from $J = 0$ to $J = 1$ (by molecules in the $J = 0$ level), or from $J = 1$ to $J = 2$ (by molecules in the $J = 1$ level), etc., are absorbed from the incoming radiation.

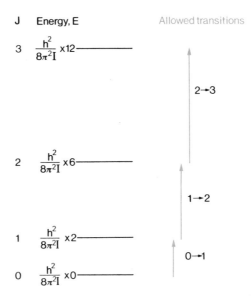

3.3 Lowest rotational energy levels for a diatomic molecule, showing the allowed transitions. J is the rotational quantum number.

For an allowed transition, $J \rightarrow J'$, the energy change involved can be written:

$$\Delta E_{rot} = \frac{h^2}{8\pi^2 I} \, [J'\,(J'+1) - J(J+1)]$$

from which it follows, since $\Delta J = 1$ (i.e. $J' = J + 1$) that:

$$\Delta E \text{ rot} = \left(\frac{h^2}{8\pi^2 I}\right) 2J' \tag{3.2}$$

where J' is the quantum number for the upper state. The energies of the transitions from the lowest levels are as follows:

Transition	Energy Change
$0 \rightarrow 1$	$\left(\dfrac{h^2}{8\pi^2 I}\right) \times 2$
$1 \rightarrow 2$	$\left(\dfrac{h^2}{8\pi^2 I}\right) \times 4$
$2 \rightarrow 3$	$\left(\dfrac{h^2}{8\pi^2 I}\right) \times 6$

This means that transitions from successive energy levels to the levels above them are associated with energy changes (ΔE) which have steadily increasing values (an increment of $2h^2/8\pi^2 I$). This is exactly what is observed in the spectrum: thus, for HF, molecules undergoing the $0{\rightarrow}1$ transition absorb energy corresponding to a wavenumber of about 40 cm^{-1}, for molecules undergoing the $1 \rightarrow 2$ transition $\bar{\nu}$ is about 80 cm^{-1}, for the $2{\rightarrow}3$ transition $\bar{\nu}$ is about 120 cm^{-1}, and so on.

(a) Calculation of the Bond Length of HF

As indicated above, the difference ($\Delta\bar{\nu}$) between two successive rotational lines is 40·5 cm^{-1}, i.e. $\Delta\nu = 1.22 \times 10^{12}$ Hz. Since $E = h\nu$, we can convert $\Delta\nu$ to the appropriate energy difference (multiplication by Planck's constant). The result must equal ($2h^2/8\pi^2 I$), as shown above. From this calculation I and r_0 can be obtained.

$$\Delta E = h\Delta\nu = \frac{2h^2}{8\pi^2 I}$$

$$I = \frac{h}{4\pi^2 \Delta\nu}$$

$$= \frac{6.626 \times 10^{-34}}{4\pi^2 \times 1.22 \times 10^{12}}$$

$$= 1.376 \times 10^{-47} \text{ kg m}^2 \text{ (or J s}^2)$$

Now, since

$$I = \left(\frac{m_1 m_2}{m_1 + m_2}\right) r_0^2 \tag{3.3}$$

where m_1 is the mass of the hydrogen atom, and m_2 is the mass of the fluorine atom, so that $m_1 = \dfrac{1}{10^3 \times L}$ kg, $m_2 = \dfrac{19}{10^3 \times L}$ kg, where L is the Avogadro Constant (6.023×10^{23} mol^{-1}), then,

$$r_0^2 = 1.376 \times 10^{-47} \times \frac{20}{19} \times 6.023 \times 10^{26}$$

$$= 0.872 \times 10^{-20} \text{ m}^2$$

$$r_0 = 0.93 \times 10^{-10} \text{ m} = 0.093 \text{ nm}$$

In this way the bond lengths (r_0) of a variety of heteronuclear diatomic molecules can be measured (see Table 3.1).

In practice, the experiments are most accurately carried out with what is known as a **microwave** (rather than a far infra-red) source of radiation (the essential theory and the range of wavelength employed is the same in the two cases, although the experimental arrangement differs somewhat). Bond lengths can then be estimated to within about 0.0005 nm. It should also be noted here that the molecules are not rigid but are actually vibrating (see later) so that the bond length measured is an *average* value. At higher rotational energy levels the average bond length also shows centrifugal distortion, and the spacings change slightly.

TABLE 3.1

Bond Lengths/nm for Diatomic Molecules determined from Rotational Spectra

HF	0.093
HCl	0.127
HBr	0.141
HI	0.160

The interpretation given above for the spacing in the rotational spectrum of a diatomic molecule is capable of being tested by employing an isotopic substitution method. Since the bond length in a molecule is effectively an *electronic* property, it should not be affected by isotopic substitution, so that, for example, ^1HF and ^2HF (deuterium fluoride) should have the same bond length. The spacing in the rotational i.r. spectrum for ^2HF is found to be approximately half that for ^1HF, which is just as expected from equations 3.2 and 3.3 if we assume that the two molecules have the same bond length.

(b) Triatomic molecules

Complications arise when a linear triatomic molecule (e.g. HCN) is investigated. A spectrum can be observed, since the molecule has a dipole moment, and the discrete energy absorptions correspond to quantised changes in the energy of rotation about an axis through the centre of gravity, perpendicular to the bonds. However, the spacing in the spectrum only leads to the appropriate moment of inertia, and this value cannot be used to derive *both* bond lengths (r_{CH} and r_{CN} for this example). The problem can be solved, however, with the help of isotopic substitution; the spacing of the energy levels, and hence the moment of inertia, is measured for ^2HCN, as well as for ^1HCN. As the bond lengths are unaltered by isotopic substitution, there is enough information (two moments of inertia, known masses) for both r_{CH} and r_{CN} to be determined.

For a more complicated molecule, an analysis to give bond lengths and angles may be possible if there is some degree of symmetry (as well as a dipole moment). For example, from a study of trichloromethane (chloroform, $CHCl_3$) the moment of inertia about an axis perpendicular to the C—H bond can be determined (rotation about an axis along this bond leads to no change in dipole moment). With the assistance of data from experiments on isotopically labelled derivatives, the following structure is established:

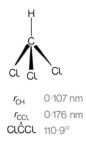

r_{CH}	0·107 nm
r_{CCl}	0·176 nm
Cl$\hat{C}$Cl	110·9°

3.2 VIBRATION–ROTATION I.R. SPECTRA OF SMALL MOLECULES

Many molecules show, in addition to the absorptions in the far i.r. or microwave region of the spectrum, characteristic absorptions in a higher energy part of the i.r. region. Most commercial spectrometers are designed to operate in this range, with both gaseous and liquid samples. As will be explained, this type of study has the advantage over the method described in Section 3.1 that extra information (besides bond lengths) can be obtained.

(a) Infra-red Spectrometer

Figure 3.4 illustrates the essential features of the instrument. The radiation is emitted from a heated filament as a continuous range of

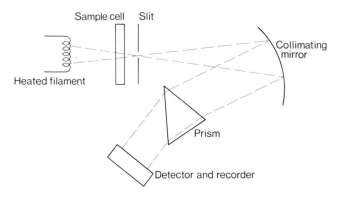

3.4   Essential features of an i.r. spectrometer

frequencies (and hence wavelengths and wavenumbers) in the i.r. region. This radiation is then passed through the sample which is contained in a cell with a path length, for a gas, of several centimetres, or, for a liquid sample, up to 10^{-2} cm. The resulting radiation, from which some of the radiation at certain frequencies will have been absorbed by molecules in the sample, is then passed through a system of slits and mirrors to emerge as a collimated beam. A prism or grating disperses the beam into components at different wavelengths (just as a prism splits up a beam of visible white light into different colours) and, depending on the orientation of the prism, radiation of separate wavelengths reaches the detector thermocouple. The detector monitors the radiation transmitted at different wavelengths and converts the radiant energy into an electrical signal. An automatic scan of frequency or wavenumber, against either energy absorbed or transmitted, is easily achieved. The prism and sample holders have to be transparent to i.r. radiation and are prepared from suitable inorganic salts (NaCl, KBr, LiF, for example). A good description of the experimental arrangement can be found in "The Physics of Chemical Structure", Revised Edition, by R. J. Taylor (*Unilever Educational Booklet*, Unilever Education Section, London, 1967).

(b) The spectrum from hydrogen chloride

Figure 3.5 shows the absorption spectrum of gaseous HCl in the wavenumber region 2600–3100 cm^{-1}; this absorption is in addition to that in the lower-energy 0–300 cm^{-1} range discussed previously. The most important features to note are the regular spacings of about 20 cm^{-1} between adjacent lines and the fact that the spectrum is centred at about 2890 cm^{-1}. Although there are separate absorptions

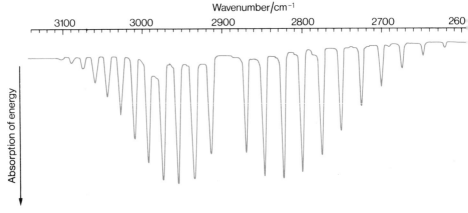

3.5   *Vibration-rotation i.r. spectrum ($\bar{v}$ 2600–3100 cm^{-1}) of gaseous hydrogen chloride,*
        HCl

from both $H^{35}Cl$ and $H^{37}Cl$ the differences in position are slight and the lines from the isotopically different species are not separated.

(c) Analysis of the spectrum

The behaviour of hydrogen chloride can be understood in terms of the molecule having vibrational energy (again, in well-defined amounts, or **quanta**) as well as rotational energy. For molecules with a dipole moment (e.g. HCl) this vibration allows interaction with the incident radiation, and hence energy can be absorbed: radiation of the appropriate energy (greater than that required for changes in the rotational energy) raises the molecule from its lowest *vibrational* energy state to the first excited vibrational state. The rotational energy can also change, which accounts for the many lines observed. This is now explained in more detail.

The quantum theory predicts that only certain vibrational energies E_{vib} are allowed. These can be expressed as follows:

$$E_{vib} = (v + \tfrac{1}{2})h\nu_0 \qquad (3.4)$$

where v is a quantum number, with possible values 0, 1, 2, etc. and ν_0 is called the **fundamental frequency.**

The lowest two energy levels, v_0 and v_1, will have $E_{vib} = \tfrac{1}{2}(h\nu_0)$ and $E_{vib} = 3/2(h\nu_0)$ respectively, so that the difference between them (i.e. the energy of the $v_0 \rightarrow v_1$ transition) is $h\nu_0$: the appropriate frequency associated with this energy change is the fundamental frequency (ν_0). It should also be noted that even in the ground vibrational state the molecule has vibrational energy, $\tfrac{1}{2}(h\nu_0)$ (this is called the zero point energy).

The molecule is rotating and vibrating simultaneously, and the *total* energy associated with these motions is the sum of the separate rotational and vibrational energies previously referred to:

$$E_{vib} + E_{rot} = (v + \tfrac{1}{2})h\nu_0 + \frac{h^2}{8\pi^2 I}J(J + 1) \qquad (3.5)$$

This means that there are many possible energy levels: a molecule in a certain vibrational energy level can still have any one of the possible rotational energy levels referred to earlier.

Allowed energy levels are illustrated in Figure 3.6: note that the vibrational energy levels are much more widely spaced than those for rotation (i.e. *ca.* 3000 cm^{-1} compared to 20 cm^{-1}—the vibrational energy separation in the Figure is not to scale).

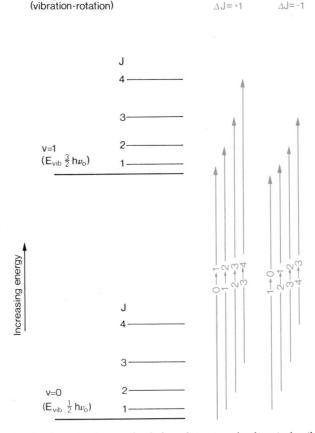

3.6   *Vibrational and rotational energy levels for a diatomic molecule: v is the vibrational quantum number and J is the rotational quantum number (N.B. the energy levels are not reproduced to scale: the vibrational levels are much more widely spaced in proportion to the separation between the rotational energy levels than are indicated here).*

When a molecule absorbs energy it can either change its J value (as discussed in Section 3.1) or it can change its vibrational *and* rotational levels. The particular transitions of the latter type which it is allowed to undergo are limited to those for which $\Delta v = +1$ (e.g. from v = 0 to v = 1, v = 1 to v = 2) *and* $\Delta J = \pm 1$. For example, a molecule in the energy level v = 0, $J = 1$ can absorb energy to be promoted either to the level v = 1, $J = 2$ *or* to v = 1, $J = 0$. When all the possibilities are considered it is found that the spectrum should consist of two series of lines—from a given J value in v = 0 to $(J + 1)$ in v = 1, and from J in v = 0 to $(J-1)$ in v = 1. The two series will have increasing energy (and hence increasing wavenumber) and decreasing energy (and wavenumber), respectively, as seen in Figure 3.6. The reader is encouraged to measure the relative ΔE values for the transitions indicated and to check that these account for the features of a typical spectrum (Figure 3.5).

The whole spectrum is centred about the transition $(v = 0) \rightarrow (v = 1)$ which has no associated change in J (i.e. $\Delta J = 0$). For the molecule HCl, this transition is forbidden, and does not take place, but the corresponding value of the energy change associated only with vibration (i.e. as if $\Delta J = 0$) can be obtained from the middle of the spectrum.

In the analysis, only transitions from $v = 0$ to $v = 1$ are indicated: this is a very reasonable approximation, since most of the molecules present will be in their lowest vibrational level rather than in higher levels ($v = 1$, 2, etc.). In contrast, as indicated by the rotational spectrum (e.g. Figure 3.2), there are molecules in a wide range of rotational levels. This situation arises because the energy increment associated with the allowed increase of vibrational energy is much larger than that for an increase of rotational energy, so that molecules have only a small probability of being in an excited vibrational state.

The analysis also explains why the individual absorptions in the rotation–vibration spectrum are approximately equally spaced. This separation is identical with that obtained for the simple rotational spectrum, so that bond lengths can be obtained as described previously (page 47).

It is helpful also to envisage this set of absorptions as centred at 2890 cm^{-1} (corresponding to the $v_0 \rightarrow v_1$ transition and to the fundamental frequency v_0). For compounds in solution the rotational "fine structure" becomes blurred and a single broad peak centred on v_0 is observed.

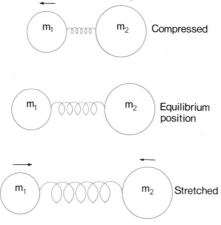

3.7   Simple model for the vibrational motion of a diatomic molecule

(d) Interpretation of the Fundamental Frequency, ν_0

The motion of the vibrating atoms which constitute a diatomic molecule is closely analogous to the simple harmonic motion of two masses attached to each other by a spring (Figure 3.7). Both systems obey Hooke's law; that is, the restoring force when the masses are stretched or compressed away from the equilibrium position is proportional to the extent of displacement from that position. For the resultant simple harmonic motion of the spring it can be shown that the frequency of oscillation, ν, is given by

$$\nu = \frac{1}{2\pi} \sqrt{\frac{k}{\mu}} \qquad (3.6)$$

where μ is the reduced mass, $m_1 m_2 / (m_1 + m_2)$, and k is the force constant of the spring (a measure of the resistance to stretching). Similarly, the behaviour of chemical bonds can be interpreted in terms of a fundamental frequency, ν_0, given by equation (3.6), where μ is the reduced mass of the two connected atoms, and, by analogy, k is the force constant of the bond. This concept is helpful in understanding vibration spectra, as the following examples show.

For $H^{35}Cl$, the absorption at wavenumber 2890 cm^{-1} in the vibrational spectrum is equivalent to a fundamental frequency, ν_0, of 8.67×10^{13} Hz. This leads to a value for the force constant k, of 4.8×10^2 N m^{-1} or kg s^{-2} (4.8×10^5 dynes cm^{-1}).

The equations for rotational and vibrational energy levels indicate that the lines for $H^{35}Cl$ and $H^{37}Cl$ should be virtually superimposed (as observed) because μ, that is $m_1 m_2/(m_1 + m_2)$, is very similar for the two molecules.

However, calculations employing these equations indicate that this similarity does not apply if we compare HCl and DCl. Thus the predicted spectrum from the latter (either chlorine isotope) will have quite different energies for its rotational and vibrational transitions when compared with the former. The force constant, an electronic property, is the same for HCl and DCl (with either chlorine isotope in each case) so that ν_0 (HCl) and ν_0 (DCl) should be related, as indicated by equation 3.6, by an amount given by their different $\sqrt{\mu}$ values: the experimental spectrum (Figure 3.8) confirms the expected ratio of $\sqrt{2}$ for ν_0 (HCl): ν_0 (DCl), with ν_0 for DCl ca. 2090 cm^{-1} (remember that wavenumber is proportional to frequency). The spectrum also confirms the expected smaller rotational splitting for DCl; the reader may like to confirm that a factor of 2 (as observed) is predicted by the theory given here. Further, the separation of the spectrum from ^2HCl (Figure 3.8) into absorptions from $^2H^{35}Cl$ and $^2H^{37}Cl$, with slightly different ν_0, is also as expected from equation 3.6 (the separation is smaller for $^1H^{35}Cl$ and $^1H^{37}Cl$).

The spectra of other heteronuclear diatomic molecules can be analysed to give force constants in similar fashion. As might be

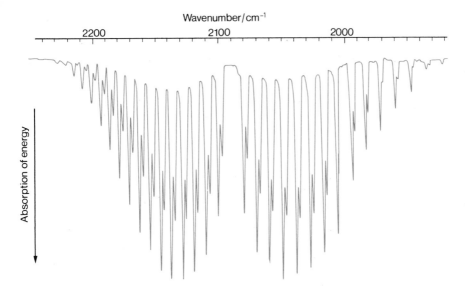

3.8   Vibration-Rotation i.r. spectrum ($\bar{\nu}$ 1900–2300 cm^{-1}) of gaseous deuterium chloride, ^2HCl (DCl)

expected, there is a correlation between force constants and bond enthalpies (energies) which is apparent, for example, in the values for the series of hydrogen halides (Table 3.2): the trend in force constants also parallels the changes in bond length (see Table 3.1). Similarly, the strong triple bond in carbon monoxide (bond enthalpy 1075 kJ mol^{-1}) has a correspondingly large force constant (18.4 $\times$ 10^2 N m^{-1}).

TABLE 3.2

Force Constants and Bond Enthalpies for some Diatomic Molecules

	Force Constant/N m^{-1}	Bond Enthalpy/kJ mol^{-1}
HF	9.7 $\times$ 10^2	562
HCl	4.8 $\times$ 10^2	431
HBr	4.1 $\times$ 10^2	366
HI	3.2 $\times$ 10^2	299

It is fortunate that another branch of spectroscopy (Raman) can be employed to calculate bond lengths and force constants for homonuclear diatomic molecules (e.g. H_2, N_2). However, since this technique does not find widespread application for organic molecules, it will not be discussed further; suitable references will be found at the end of this chapter.

(e) Triatomic Molecules

For molecules more complicated than the diatomic molecules considered so far, there are several possible ways in which the bonds can vibrate (these are called vibrational **modes**) and each vibration has an associated fundamental frequency. For example, the fundamental modes for the linear molecule CO_2 are illustrated in Figure 3.9. Of these, ν_1 (the *symmetric stretching* mode) does not involve a dipole moment change and is "inactive" in the infra-red region, there being no corresponding absorption of energy. However, absorption peaks

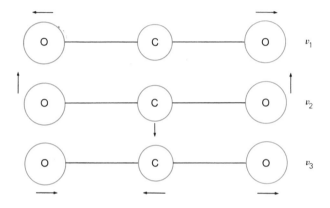

3.9   Vibrational modes for carbon dioxide, CO_2

are observed for ν_2 (667 cm^{-1}) and ν_3 (2349 cm^{-1}) because these modes of vibration involve dipole moment changes. These are called *bending* (ν_2) and *asymmetric stretching* (ν_3) modes, respectively, and the numerical values indicate (as might be expected) that less energy is involved in bending than in stretching.*

For a non-linear triatomic molecule (e.g. H_2O) the three vibrational modes are all infra-red active since each involves a dipole moment change. Since three absorptions are detected in the vibrational i.r. spectrum of SO_2, it can be concluded that this molecule (like H_2O) is not linear.

---

* A fairly simple apparatus for demonstrating fundamental vibrational modes for ball-and-spring "molecules" is described by D. Luke, "A mechanical analogue for modes of vibration in the infra-red", *The School Science Review*, 1972, **54**, 105.

3.3 I.R. SPECTROSCOPY OF ORGANIC MOLECULES

The i.r. absorption spectrum gets very rapidly more complicated as we progress from diatomic molecules to triatomic, tetra-atomic and larger molecules, because of the increase in the number of possible vibrations. Only in fairly simple cases can a full analysis be carried out, but nevertheless it is often possible to obtain useful information about the structure of complex molecules from i.r. spectra.

The compound under investigation is examined, where possible, as a liquid sample. This can be done for a neat liquid by squeezing a few drops between two KBr discs (these are transparent to i.r. radiation) or, if the compound is a solid, by grinding it up to form a "mull" with paraffin oil before placing it between the discs. Alternatively, the compound to be studied may be dissolved in a suitable solvent (e.g. $CHCl_3$, CCl_4); although peaks characteristic of the solvent's i.r. absorptions will be observed, these can easily be recognised and allowance made for them. (In a double beam spectrometer, peaks from the pure solvent are recorded simultaneously and subtracted automatically.) The spectra are plots of **transmittance** of energy at a given wavelength, rather than absorption of energy. Transmittance and absorption are related: a large transmittance implies little absorption, and vice-versa.

Organic compounds show spectra in which many peaks are spread over the wide scan-range customarily employed ($5000-650$ cm^{-1}). Each peak is associated with a particular vibration (or a combination of these)—the rotational fine structure is smeared out for molecules in the liquid phase because rotation of the molecules is not free, as it is in a gas; each peak is really an "envelope" of all rotational lines. The complexity of the spectra (see, for example, the i.r. spectrum of propanone, Figure 3.10) reflects the large number of fundamental

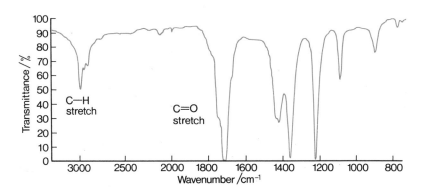

3.10   I.r. spectrum of propanone, $(CH_3)_2CO$ (liquid film)

vibrations, which depends on the complexity of the molecule (i.e. its number of bonds). Fortunately, we can to a certain extent associate certain absorptions with stretching or bending vibrations of particular bonds (or sometimes groups) in a molecule: for example, the absorptions at *ca.* 3000 and 1700 cm^{-1} in the spectrum of propanone are typical of the stretching modes of C—H and C=O groups, respectively, in organic compounds. In this way the spectra allow recognition of the types of functional group present in an organic molecule.

Other absorptions are observed which are characteristic of the molecule as a whole, these depending on interactions between different groups. Further, there are complications which arise because absorptions can occur at "overtones" (harmonics) of other frequencies or at "combinations" ($v_1 + v_2$) and "differences" ($v_1 - v_2$) of other frequencies. These all add to the complexity but do provide the equivalent of an unambiguous finger-print for any particular molecule.

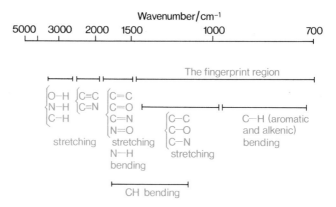

3.11 Typical areas for absorptions in the i.r. spectra of organic molecules

Dependence of the i.r. spectrum on molecular structure

Typical absorptions are indicated diagrammatically in Figure 3.11 and this information can be of considerable assistance in the determination of an unknown compound's structure. The important features of the Figure can be helpfully interpreted in terms of the dependence of a characteristic bond stretching frequency (and hence wavenumber) on the *force constant* for that bond and on the *masses* of the atoms which are joined by the bond.

For example, the lowering of the wavenumber (and hence energy) for stretching in the series *triple bond* > *double bond* > *single bond* [for C—C and C—N bonds and, if carbon monoxide is included (C≡O, $\bar{v}_0$ 2146 cm^{-1}), for C—O bonds], is as expected from the appropriate bond enthalpies (bond strengths) and hence force constants (see equation 3.6). Further, variations in the reduced mass (μ) of the atoms forming the bond (see equation 3.6) also has predictable effects, notably the high wavenumber (and hence frequency) for C—H, N—H and O—H stretches (these have low values of μ). Lastly, we can note that since bending involves less energy than stretching, bending absorptions occur at lower wavenumbers than stretching modes involving the same bonds.

Table 3.3 lists approximate force constants for some bonds in organic molecules. They are derived from the corresponding stretching vibrational frequencies and equation 3.6, page 54, using the masses of the atoms concerned (i.e. C and H for the C—H stretch which occurs at *ca.* 3000 cm^{-1} for most organic molecules). They are fairly typical values for the kind of bond indicated and are appropriate for a variety of molecules.

TABLE 3.3

Typical Force Constants and Average Bond Enthalpies for Bonds in Organic Molecules

	Force Constant/$N\ m^{-1}$	Average Bond Enthalpy/$kJ\ mol^{-1}$
C—H (alkanes)	4.8×10^2	414
C—C (ethane)	4.6×10^2	347
C=C (ethene)	10.8×10^2	611
C≡C (ethyne)	14.9×10^2	837
C≡N (ethanonitrile)	17.3×10^2	892

One of the particular attractions of i.r. spectroscopy is that a more detailed investigation of the positions of absorption of a group (e.g. C—H) in a variety of different molecules shows that the exact wavenumber does depend to a small but significant extent on the environment of that group in a molecule (cf. chemical shifts in n.m.r. spectra, Chapter 5) and this can prove diagnostically extremely valuable. Some characteristic absorptions which illustrate the variation are listed in Table 3.4 and these are now discussed in more detail.

(i) *C—H bonds.* For example, methyl (CH_3) and methene (CH_2) groups usually exhibit C—H stretching modes in the range 2950–2850 cm^{-1} (sometimes these appear with splittings because of interaction between the different C—H bonds) and also bending modes at around 1450 cm^{-1}. The absorptions from the C—H stretch in propanone at just below 3000 cm^{-1} and the bending modes at *ca.* 1400 cm^{-1} are clearly visible in Figure 3.10.

In contrast, an alkanal C—H stretch usually appears between 2700 and 2900 cm^{-1}, and other C—H stretching absorptions are as follows: alkynes (C≡CH), 3300 cm^{-1}; alkenes (C=CH$_2$), 3095–3075 cm^{-1}; arenes 3040–3010 cm^{-1}. Out-of-plane bending for arene and alkenic hydrogen atoms often gives characteristic absorption bands at 900–650 and 980 cm^{-1}, respectively.

The characteristic arene C—H stretching vibrations (> 3000 cm^{-1}) and out-of-plane bending vibration (680 cm^{-1}) are apparent in the i.r. spectrum of benzene (Figure 3.12). Other prominent absorptions in this example are from in-plane C—H bending (1040 cm^{-1}) and C—C stretching (1480 cm^{-1}). For more complicated benzene derivatives, it is sometimes possible to determine the ring substitution pattern from the modifications produced in the i.r. spectrum.

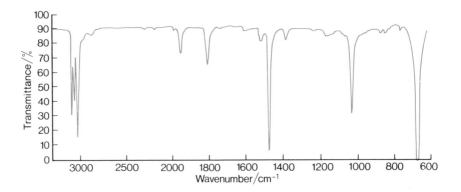

3.12   I.r. spectrum of benzene, C$_6$H$_6$ (liquid film)

(ii) *O—H, C—O bonds*. Hydroxyl groups (OH) generally give a strong peak (from the stretching vibration) in the 3650–3590 cm^{-1} region; if the group is hydrogen-bonded there is sometimes a broadening and a shift towards lower wavenumbers.

For example, Figure 3.13 is the i.r. spectrum of ethanol, showing clearly the C—H and O—H stretches (2900 and 3300 cm^{-1}, respectively) and absorption from aliphatic C—H bending (1400 cm^{-1}) and C—O stretching (1050 cm^{-1}, see also section (iv)). From the shape and position of the O—H peak it can be concluded that this group is taking part in hydrogen-bonding. Other spectroscopic manifestations of hydrogen-bonding will be encountered in the next two chapters.

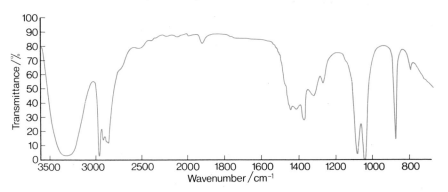

3.13   I.r. spectrum of ethanol, CH₃CH₂OH (liquid film)

   (iii) *N—H bonds*. Like O—H bonds, N—H bonds show a characteristic absorption at high wavenumber (3500–3300 cm⁻¹). Hydrogen-
bonding involving the N—H hydrogen may again cause broadening
and a shift towards lower wavenumbers.

   (iv) *C═O bonds*. Carbonyl-containing compounds show a very
characteristic strong C═O absorption in the range 1800–1600 cm⁻¹,
the exact value of which depends on the structure of the adjacent
groups. For simple ketones and for aliphatic alkanals the appropriate
value is 1725 cm⁻¹ (see Figure 3.10) but the wavenumber tends to be
slightly higher for simple alkanoyl chlorides, esters, and anhydrides,
and slightly lower than this for amides. The characteristic wavenumbers
are also generally a little lower if the carbonyl group (in ketones, esters,
etc.) is adjacent to an alkenic double bond (C═C—C═O) or to a
phenyl ring or if it is involved in hydrogen-bonding (see also Chapters
4 and 5).

   Carbon—oxygen single bonds (C—O) in esters, alkanols, and
ethers, usually show a strong absorption in the range 1300–1050 cm⁻¹
[see section (ii)].

   It should here be stressed that the absorptions mentioned in this
section and listed in Table 3.4 are only some of the characteristic
vibrations which may be encountered—the observed spectra often
contain complicating features from other parts of the molecule.
However, the information given here on the dominant peaks and
commonly occurring groups should allow definite conclusions to be
drawn about the molecular structure of unknown compounds.

TABLE 3.4

Characteristic Infra-red Absorptions for a variety of Organic Molecules*

Molecule or Group	Vibration type	Wavenumber/cm^{-1}
Alkyl group (CH_3, CH_2, CH)	{ C—H stretch { C—H bend	2960–2850 1460–1370
Alkanal (CHO)	C—H stretch	2900–2700
Alkyne (C≡CH)	C—H stretch	3300–3270
Alkene (C=CH_2)	{ C—H stretch { C—H bend	3095–3075 990– 890†
Arene	⎰ C—H stretch ⎨ C—H bend: in-plane ⎱ out-of-plane	3040–3010 1300–1000 900– 650†
Alkanol (OH)	{ O—H stretch { C—O stretch	3650–3590‡ 1200–1050
Amine, amide (NH_2)	N—H stretch	3500–3300‡
Aliphatic ketone (R_2CO)	C=O stretch	1740–1700
Aliphatic alkanal (RCHO)	C=O stretch	1740–1720
Aromatic ketone (Ar_2CO)	C=O stretch	1700–1680
Alkanoic acid (RCO_2H)	C=O stretch	1725–1700
Alkanoyl chloride (RCOCl)	C=O stretch	1815–1790
Alkanoate ester (RCO_2R')	{ C=O stretch { C—O stretch	1750–1730 1300–1050
Alkoxy (ether) R_2O	C—O stretch	1150–1070

*Some of these peaks may be split into several components; see p. 59.
† Characteristic variations occur with different substitution patterns.
‡ These may be drastically affected by hydrogen-bonding; see pp. 60, 61.

3.4 EXAMPLES OF INFRA-RED SPECTRA

The following examples illustrate some of the features outlined above and indicate the variation of i.r. spectra with molecular structure.

(a) Hexane, $CH_3(CH_2)_4CH_3$ (Figure 3.14)

This spectrum reveals very clearly the characteristic aliphatic C—H stretching modes (just below 3000 cm^{-1}) and bending modes (1500 cm^{-1}), with no indication of any functional groups.

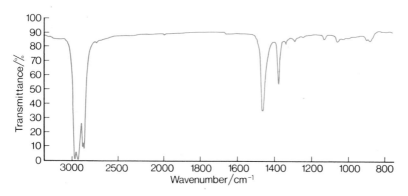

3.14   I.r. spectrum of hexane, $CH_3(CH_2)_4CH_3$ (liquid film)

(b) Pent-1-ene, $CH_3(CH_2)_2CH=CH_2$ (Figure 3.15)

The peaks at 900 cm^{-1} and 3080 cm^{-1} provide conclusive evidence for the presence of an alkene group: these are the alkenic C—H out-of-plane bending and stretching modes, respectively. Also present are alkyl C—H modes (*ca.* 2900 and 1450 cm^{-1}) and the C=C stretching vibration (1640 cm^{-1}).

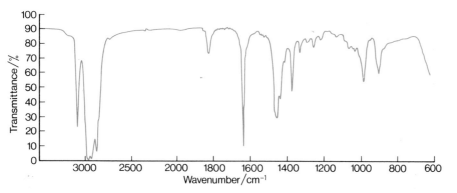

3.15   I.r. spectrum of pent-1-ene, $CH_3(CH_2)_2CH=CH_2$ (liquid film)

Alkenes without a terminal double bond show slightly different absorptions below 1000 cm^{-1}: the details of the absorptions usually allow the substitution pattern of the alkene to be established (so that, for example, *cis* and *trans* alkenes can be distinguished).

(c) Methylbenzene, $C_6H_5CH_3$ (Figure 3.16)

In addition to the aromatic and aliphatic C—H stretching modes (at 3050 and 2900 cm^{-1}, respectively) this spectrum has characteristic peaks from the aliphatic C—H bending modes (*ca.* 1500 cm^{-1}) and from out-of-plane aromatic C—H bending (*ca.* 700 cm^{-1}).

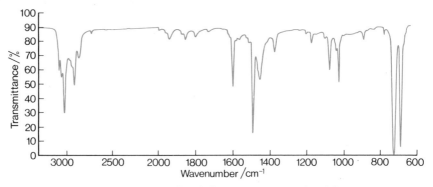

3.16   *I.r. spectrum of methylbenzene, $C_6H_5CH_3$ (liquid film)*

The particular pattern around 700 cm^{-1} is typical of a mono-substituted benzene derivative.

(d) Ethanoic acid (acetic acid), CH_3CO_2H (Figure 3.17)

This spectrum, recorded for a thin film of the neat liquid, shows typical C—H bending vibrations (*ca.* 1400 cm^{-1}) and the characteristic carbonyl absorption at 1720 cm^{-1}.

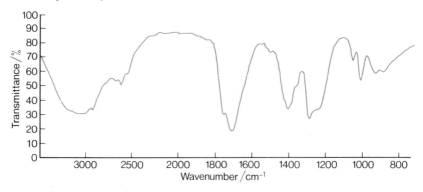

3.17   *I.r. spectrum of ethanoic acid, CH_3CO_2H (liquid film)*

In particular, the broad absorption at > 3000 cm⁻¹ typifies a hydrogen-bonded OH group.

(e) Ethyl ethanoate (ethyl acetate), $CH_3CO_2C_2H_5$ (Figure 3.18)

This shows clearly the typical carbonyl stretching absorption at 1740 cm⁻¹; note that this is at slightly higher wavenumber than that observed for ketones.

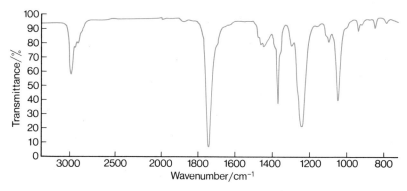

3.18   *I.r. spectrum of ethyl ethanoate, $CH_3CO_2CH_2CH_3$ (liquid film)*

Another characteristic feature is the absorption at 1240 cm⁻¹; this is a C—O stretching frequency and is particularly prominent for esters.

(f) Diethylamine, $(C_2H_5)_2NH$ (Figure 3.19)

The broad peak at *ca.* 3300 cm⁻¹ in this spectrum is indicative of an N—H group; the broadening suggests that the hydrogen atom takes part in intermolecular hydrogen-bonding with nitrogen atoms in other molecules. The strong absorption at 1140 cm⁻¹ is from the C—N stretching vibration (cf. C—O stretch in alcohols, ethers, etc.).

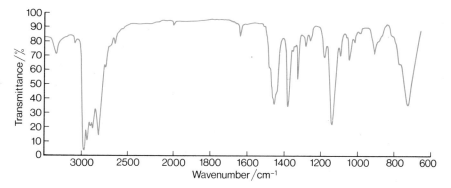

3.19   *I.r. spectrum of diethylamine, $(CH_3CH_2)_2NH$ (liquid film)*

3.5 PROBLEMS

3.1 Calculate the fundamental stretching frequency and wavenumber for a C—H bond, given that the atoms vibrate independently of other groups on the carbon atom and that the force constant is $4.8 \times 10^2 \text{ N m}^{-1}$.

3.2 Calculate the separation (in wavenumbers) between the rotational transitions in the far infra-red (or microwave) spectrum of carbon monoxide, given that the bond length, r_{CO} is 0.113 nm.

3.3 Account for the observation that in the i.r. spectrum of deuterated benzene ($C_6{}^2H_6$), Figure 3.20, the C—D (C—^{2}H) symmetric stretching mode occurs at *ca.* 2280 cm^{-1} (in undeuterated benzene the corresponding absorption is at 3050 cm^{-1}).

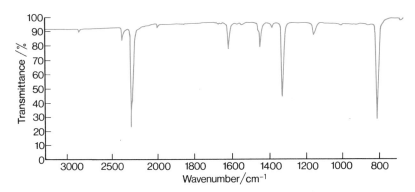

3.20   I.r. spectrum of hexa-^{2}H-benzene (hexadeuterobenzene), $C_6{}^2H_6$ (liquid film)

3.4 Figure 3.21 is the i.r. spectrum of the compound whose mass spectrum has been given previously (Problem 1.3) and whose n.m.r. spectrum is shown on page 114. Check that the i.r. spectrum supports your previous assignment.

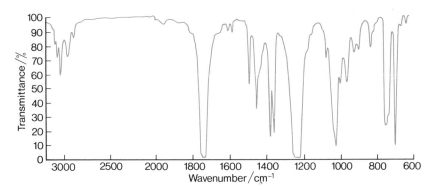

3.21   I.r. spectrum of unknown compound; Problem 3.4

More problems involving i.r. spectra can be found at the end of the book (p. 148).

3.6 CONCLUSION

The usefulness of i.r. spectroscopy lies not only in the detailed information which can be obtained for small molecules, but also, for organic molecules, in the rapid and fairly inexpensive technique it provides for deciding which functional groups are present in a molecule. It has some advantages over mass spectra (Chapter 1) and n.m.r. (which is discussed in Chapter 5); thus, compared with the former it has the advantage that it can be easily applied to non-volatile samples. Compared with n.m.r. it possesses the ability to give information about all the atoms in a molecule (not just those nuclei with magnetic moments). On the other hand, for organic compounds the i.r. spectra are, on the whole, not so rich in detailed information as the n.m.r. or mass spectra. In practice all three techniques (and also ultra-violet spectroscopy, Chapter 4) can usually be used to complement each other in a complete diagnosis. Thus details of the environment of a C=O group, for instance, as indicated by the i.r. spectra of amides, ketones, esters, etc., are often complementary to the details which n.m.r. reveals about the environment of hydrogen atoms in organic compounds.

Further, i.r. spectroscopy does find applications in a wider context. For example, Figure 3.22 shows the i.r. spectrum obtained from a thin film of nylon-6.6 (this polymer is prepared by the **copolymerisation** reaction of hexanedioic acid and hexane-1,6-diamine). Infra-red absorptions from C—H stretching (2900 cm^{-1}), amide C=O stretching

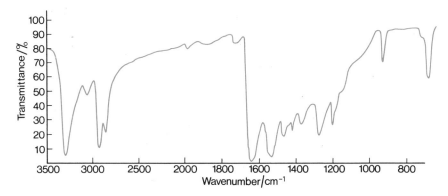

3.22   I.r. spectrum of nylon-6.6 (thin film)

(1660 cm⁻¹) and N—H stretching (3300 cm⁻¹) help confirm the following structure:

$$\left[\begin{array}{c} \overset{O}{\underset{\|}{C}}CH_2CH_2CH_2CH_2\overset{O}{\underset{\|}{C}}N\underset{H}{|}CH_2CH_2CH_2CH_2CH_2CH_2N\underset{H}{|} \end{array}\right]_n$$

The bonds around the amide function (the **peptide link** CO—NH) are planar; out-of-plane N—H bending accounts for the absorption at *ca.* 700 cm⁻¹.

A thin film of perspex has strong absorptions at *ca.* 2950, 1730, 1450 and 1200 cm⁻¹. What type of polymer structure is consistent with these observations? A quite different application is provided by a structural investigation of the metal–carbonyl compound $Fe_2(CO)_9$, (Figure 3.23), which has i.r. absorptions at 2020 and 1830 cm⁻¹. The former is from terminal carbonyl groups (rather similar to carbon monoxide itself, which has $\bar{v}$ 2146 cm⁻¹) whereas the latter is from the bridging carbonyl groups (cf. $R_2C{=}O$, *ca.* 1700 cm⁻¹). This information, when considered with the molecular formula, provides confirmatory evidence for the structure.

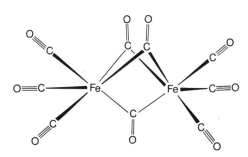

3.23   Structure of $Fe_2(CO)_9$

Further reading

G. C. Pimentel, "Infra-red Spectroscopy: a Chemist's Tool", *J. Chemical Education,* 1960, **37,** 651.

W. B. Simpson, "Some Notes on the Applications of Infra-red Spectroscopy to Inorganic Chemistry", *Education in Chemistry,* 1966, **3,** 58.

Advanced reading (theory of i.r. and Raman spectroscopy)

P. J. Wheatley, "The Determination of Molecular Structure", 2nd Edn., Oxford University Press, 1968.

G. M. Barrow, "The Structure of Molecules", Benjamin, New York, 1964.

Advanced reading (structure determination)

J. R. Dyer, "Applications of Absorption Spectroscopy of Organic Compounds", Prentice-Hall, Englewood Cliffs, 1965.

R. M. Silverstein, C. G. Bassler, and T. C. Morrill, "Spectrometric Identification of Organic Compounds", 3rd Edn., Wiley, New York, 1974.

D. H. Williams and I. Fleming, "Spectroscopic Methods in Organic Chemistry", 2nd Edn., McGraw-Hill, London, 1973.

S. F. Dyke, A. J. Floyd, M. Sainsbury, and R. S. Theobald, "Organic Spectroscopy: an Introduction", Penguin, London, 1971.

Film

"Infra-red Spectroscopy", Perkin-Elmer Ltd (on free loan).

Chapter 4

Electronic Absorption Spectroscopy

In addition to the absorption of well-defined amounts of energy to increase its vibrational and rotational energy, a molecule may also absorb energy to increase the energy of its electrons. The energy changes involved are considerably greater than those involved in vibrational and rotational energy changes and correspond to radiation in the *ultra-violet* (λ *ca.* 200–400 nm) and *visible* (λ 400–750 nm) regions of the electromagnetic spectrum.

For radiation in the middle of this range ($\lambda = 400$ nm, $\nu = 7.5 \times 10^{14}$ Hz) the energy of one quantum ($h\nu$) is 5×10^{-19} J. This is the energy which is absorbed by *one molecule* if it absorbs one quantum of violet light, and it is obviously a very small quantity. However, for one *mole* of material (in which the number of molecules is the Avogadro constant, L) the total energy absorbed (L quanta) corresponds to $(5 \times 10^{-19}) \times (6 \times 10^{23})$ i.e. *ca.* 300 kJ: this is the same order of magnitude as some *bond enthalpies* (energies) in typical molecules. It is therefore not surprising that the interaction of visible or ultra-violet radiation with molecules can sometimes bring about chemical reactions involving bond breakage: such processes are called **photochemical** reactions.

4.1 ELECTRONIC ENERGY CHANGES

Figure 4.1 represents diagrammatically the ground electronic state (E_0) of a molecule, together with its associated vibrational and rotational energy levels. The overall energy of the molecule is the sum of contributions from electronic, vibrational and rotational energy, i.e.

$$E = E_{\text{elec}} + E_{\text{vib}} + E_{\text{rot}}$$

where E_{vib} and E_{rot} have exactly the same allowed values discussed previously (Chapter 3).

In an ordinary sample, all the molecules will be in the E_0 level but may have different v and J quantum numbers. When the appropriate energy is incident it is possible to excite the molecule to the higher energy *electronic* state E_1 (which again has similarly quantised rotational and vibrational energy levels). The excited state arises because of the promotion of electrons to higher energy **molecular orbitals,** exactly as can happen for electrons in atomic orbitals (cf. the sodium flame test, page 41).

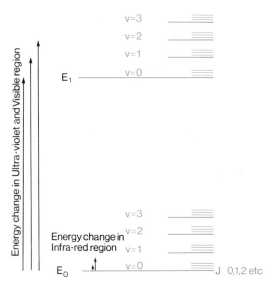

4.1   Diagrammatic representation of electronic, vibrational, and rotational energy levels

It is this type of absorption which explains why some compounds are coloured—the absorption of energy for the transition $E_0 \to E_1$ is in the visible region. Other compounds absorb energy in the ultra-violet region and although they do not appear coloured (unless there is also visible absorption) the absorption in this part of the spectrum can be detected. It is this absorption of energy with which we will here be concerned. Although it is beyond the scope of this book, it is worth noting that the phenomena of fluorescence and phosphorescence are associated with the re-emission of energy (particularly in the visible region) when the molecules in an excited electronic state return to the ground state.

We will return to a further discussion of molecular orbitals and excited electronic states when some examples of electronic spectra have been considered.

4.2   ELECTRONIC   ABSORPTION   SPECTROSCOPY   OF ORGANIC MOLECULES

(a) The Experiment

The experimental method is similar to that employed for infra-red spectroscopy, and many commercial instruments are available. The source provides radiation in a continuous range of wavelengths in the u.v. and visible region, as from a heated tungsten filament (e.g. an electric light bulb). In practice, two separate sources are usually required for covering the whole u.v. and visible region. A prism separates the radiation into component wavelengths and the absorption of the sample at any particular wavelength is measured by the reduction of the signal from a photoelectric device when the sample is placed in the incident beam.

A simple example is provided by the vapour of molecular iodine whose spectrum consists of a large number of closely spaced absorptions in the $\lambda = 550$ nm region (absorptions are traditionally characterised by wavelength in the u.v.-visible part of the spectrum). The various transitions are from the lowest vibrational states of the ground electronic state to excited vibrational states in E_1 (see Figure 4.1). The resulting lines are called vibrational fine-structure (cf. rotational fine-structure in i.r. spectra): more details of the spectrum are given in a reference at the end of this chapter. The $E_0 \rightarrow E_1$ absorptions at about 550 nm account for the colour of the vapour.

Most investigations involve the use of liquid samples, which usually contain the compounds to be studied as solutions in suitable solvents. Modern spectrometers employ a *double-beam* system whereby two identical beams of radiation are generated, one of which passes through the solution under investigation, the other of which passes through an equivalent amount of pure solvent, so that the difference in absorption which is measured is just that due to the molecules of the *solute* in the solution; the solvent should be transparent in the region of interest. Tetrachloromethane, hexane, cyclohexane and ethanol are often employed as solvents, and the solutions are usually contained in glass cells such that the beam of radiation passes through 1 cm of solution (this is known as the path length).

Some spectrometers automatically record absorption as a function of wavelength. In others the operator records the absorption at different wavelengths. A colorimeter is a simple type of spectrometer working with a given colour (i.e. at a fixed range of wavelength in the visible region, selected with a suitable filter from a wide range of radiation).

(b) Examples of spectra

Figures 4.2–4.4 are the electronic absorption spectra of three organic molecules—propanone, benzene and the indicator methyl red. The spectra are plots of wavelength (λ/nm) against absorbance (A) (which is sometimes also referred to as optical density). The absorbance is the logarithm of the ratio of the intensity of the incident radiation (I_0) to that of the transmitted radiation (I):

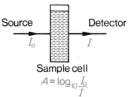

$$A = \log_{10} \frac{I_0}{I}$$

A peak in the spectrum at a given λ corresponds to absorption of energy at this wavelength by the solute molecules; for some molecules, more than one area of absorption is observed, as in Figures 4.2 and 4.3 (strong absorptions tailing off into the normally inaccessible region below 200 nm are detected in addition to the higher-wavelength peaks). Propanone and benzene are colourless, since they do not absorb in the visible region, but methyl red absorbs in acid solution in the blue-green region (λ 400–600 nm) and so appears red; in alkaline solution, the absorption is at lower wavelengths and the solution is yellow.

The peaks observed are broad in most cases because molecular interactions in the liquid cause the obliteration of the expected vibrational and rotational fine structure (although vibrational fine-structure is clearly apparent in the peak with λ 255 nm in the spectrum from benzene).

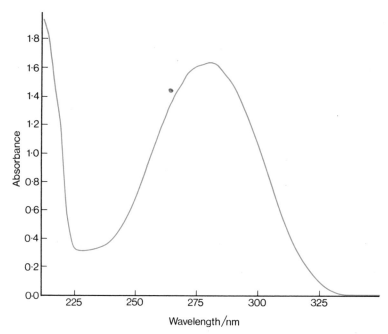

4.2   *Electronic absorption spectrum of propanone,* $(CH_3)_2C{=}O$ *(recorded for a solution of propanone in hexane, of concentration 6 g dm^{-3}, using a cell with a path length of 1 cm).*

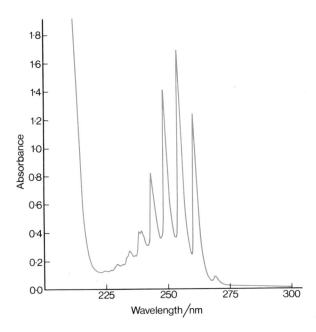

4.3 *Electronic absorption spectrum of benzene, C_6H_6 (recorded for a solution of benzene in hexane, of concentration 0.6 g dm^{-3}, using a 1 cm cell).*

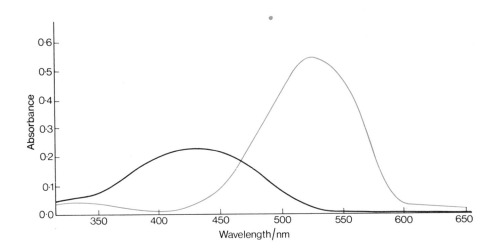

4.4 *Electronic absorption spectra of aqueous solutions of the indicator methyl red (both containing 0.004 g dm^{-3} of the indicator; —— pH 1, —— pH 13; recorded with a 1 cm cell).*

Electronic absorption spectra are usually characterised by two parameters

(i) the values of the *wavelengths* at which *absorption maxima* occur (λ_{max}). Inspection of Figures 4.2–4.4 indicates the different values obtained for the different molecules considered. For example, for propanone, the high-wavelength absorption has $\lambda_{max} = 279$ nm

(ii) the *extent* of absorption, for a given concentration of compound at any given wavelength (i.e. the *height* of the peak): this is the subject of the next section.

As will be seen, the position (λ_{max}) and extent of absorption provide two more characteristic properties of a molecule which depend on its structure.

(c) The Beer–Lambert Law

The extent of absorption at a given wavelength by an absorbing compound in a non-absorbing solvent is found to depend upon the *concentration* of the compound and upon the *path length* of the cell. The **Beer–Lambert Law**, which is generally well-obeyed for fairly dilute solutions, expresses the dependence of the absorbance on these two variables:

$$A = \log_{10}\frac{I_0}{I} = \varepsilon c b \qquad (4.2)$$

where A is the absorbance
       b is the path length
       c is the concentration
       ε is a constant for a particular compound at a chosen wavelength.

If b is in m and c is in mol m^{-3}, then ε, in m^2 mol^{-1}, is described as the **molar decadic absorptivity** or **molar extinction coefficient.** Thus if b and c are known, and if the experiment gives a value for A (the spectrometer records this), ε can be calculated. This is usually quoted for the wavelength of maximum absorption (i.e. at λ_{max}).

It is important to realise that, once ε has been determined, the value of A for a given solution of a known compound can be used to determine c, the concentration of that compound in the solution. This behaviour, the basis of the Beer–Lambert Law, means that u.v.-visible spectrometry provides an excellent means for quantitative, as well as qualitative analysis.

4.3 THE RELATIONSHIP OF λ_{max} AND ε_{max} TO STRUCTURE

Investigation of the electronic absorption spectra of a variety of organic compounds demonstrates that only certain types of molecule exhibit absorption in the u.v.-visible range (λ 200–750 nm). The feature that these species have in common is that they contain double or triple bonds (or, in some cases, lone-pairs of electrons), which are essentially responsible for the absorption; these fragments are called **chromophores**. It is also found that when two or more chromophores are adjacent to each other (the groups are then said to be **conjugated**) the absorptions become more pronounced (higher ε_{max}) and occur at lower energy (greater λ_{max}).

The reasons for these trends can be understood in terms of the types of molecular orbitals involved in the electronic excitations. There are three types of molecular orbital, and the electrons in these orbitals have somewhat different environments. First, there are the electrons in the σ-orbitals which constitute the bonding framework of a molecule (e.g. the electrons in the 4 C—H bonds in methane, CH_4, are described as being in σ-molecular orbitals, each of which contains two electrons). These orbitals are formed from atomic s, sp, sp^2 and sp^3 orbitals. Second, there are electrons in π-orbitals, which are formed from laterally overlapping atomic p-orbitals in compounds such as benzene and ethene (ethylene). Third, there are *lone-pair* electrons in orbitals on atoms like oxygen, nitrogen, etc.; these are called non-bonding, or n-electrons. The carbonyl group (in a ketone, say) contains all three types of molecular orbital:

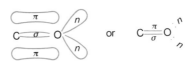

When two electrons in atomic orbitals are brought together to form a bond (i.e. when a filled molecular orbital is produced) a higher energy **anti-bonding** orbital is also formed which is empty in the ground state of the molecule (cf. empty high-energy atomic orbitals). When excitation takes place, an electron from one of the filled orbitals (σ, π, or n) becomes excited to a vacant anti-bonding orbital (σ^*, π^*) so that a new *excited* state is reached. Since various excitations are possible, there are various possible absorptions, corresponding to the transitions $n \rightarrow \sigma^*$, $\sigma \rightarrow \pi^*$, etc.

The *approximate* relative energies of typical σ, π, n and anti-bonding orbitals are indicated in Figure 4.5; as expected, the σ-electrons are the most tightly bound (most energy is needed to excite them). The order of decreasing energy for the absorptions is as follows:

$$\sigma \rightarrow \sigma^* > \sigma \rightarrow \pi^* \sim \pi \rightarrow \sigma^* > \pi \rightarrow \pi^* \sim n \rightarrow \sigma^* > n \rightarrow \pi^*$$

Of all these possible transitions, only those of the last three types normally account for absorption in the u.v.-visible region, the others being of too great an energy. This then explains why only molecules with n or π electrons give rise to characteristic u.v. and visible spectra whereas alkanes, for example, show no absorption.

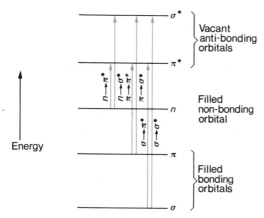

4.5   *Approximate relative energies for electrons in different types of molecular orbital in organic compounds (not to scale)*

Ethanol absorbs radiation of wavelength *ca.* 200 nm and below (it is transparent above this wavelength and actually finds use as a solvent for u.v.-visible studies on molecules with higher-wavelength absorption). This absorption derives from the $n \rightarrow \sigma^*$ absorption by the lone-pair electrons on the oxygen atom.

Table 4.1 indicates the measured values of ε_{max} and λ_{max}, together with the types of transition, for a variety of organic molecules which contain chromophores (double bonds etc.).

For propanone (see Figure 4.2) the peak at $\lambda = 188$ nm (for which $\varepsilon = 90$ m^2 mol^{-1}) is responsible for absorption at the low-wavelength end of the observed spectrum, and the peak at $\lambda = 279$ nm ($\varepsilon = 1.5$ m^2 mol^{-1}) is also clearly visible. These are due to transitions involving the electrons in the carbonyl-group double bond ($\pi \rightarrow \pi^*$) and the oxygen's lone-pair electrons ($n \rightarrow \pi^*$), respectively. The characteristic values are slightly sensitive to the solvent used.

These absorptions do not depend to any marked extent on the nature of the alkyl groups attached to the carbonyl function; thus for a variety of ketones and alkanals, the u.v. spectra recorded for hexane solutions show the following values of λ and ε for the $n \rightarrow \pi^*$ absorption: butanone (279 nm, 1.6 m^2 mol^{-1}), cyclohexanone (285 nm, 1.4 m^2 mol^{-1}), ethanal (acetaldehyde, 293 nm, 1.2 m^2 mol^{-1}), propanal (290 nm, 1.8 m^2 mol^{-1}). For carbonyl-containing compounds of different chemical type (e.g., alkanoate esters), the absorptions are, however, characteristically different (see Table 4.1).

TABLE 4.1

Compound	Solvent	λ_{max}/ nm	Transition type	ε_{max}/ m^2 m
a) Molecules with single chromophores:				
Propanone CH$_3$CCH$_3$ (with O double bond)	hexane	188 / 279	$\pi \rightarrow \pi^*$ / $n \rightarrow \pi^*$	90 / 1·5
Ethyl ethanoate CH$_3$C(=O)OCH$_2$CH$_3$ (ethyl acetate)	water	204	$n \rightarrow \pi^*$	6·0
Pent-1-ene CH$_3$CH$_2$CH$_2$CH=CH$_2$	hexane	190	$\pi \rightarrow \pi^*$	1000
Nitromethane CH$_3$N(=O)O$^-$	hexane	278	$n \rightarrow \pi^*$	1·7
(b) Conjugated molecules:				
Buta-1,3-diene, CH$_2$=CH—CH=CH$_2$	hexane	217	$\pi \rightarrow \pi^*$	2 100
Butenone CH$_2$=CH—C(=O)—CH$_3$	ethanol	219 / 324	$\pi \rightarrow \pi^*$ / $n \rightarrow \pi^*$	360 / 2·4
Benzene	hexane	184 / 203 / 255	all $\pi \rightarrow \pi^*$	6000 / 740 / 20
Phenylethanone (acetophenone)	ethanol	199 / 246 / 279 / 320	all $\pi \rightarrow \pi^*$	2 000 / 1 260 / 100 / 4·5
Nitrobenzene	hexane	252 / 280 / 330	all $\pi \rightarrow \pi^*$	1000 / 100 / 12·5

When a carbon—carbon double bond is present, a characteristic $\pi \rightarrow \pi^*$ absorption is observed (Table 4.1 gives data for pent-1-ene): what is particularly noticeable here is the larger value of ε_{max} (which means that a more dilute solution of the compound is needed, compared with the ketones, to obtain the same amount of absorption). This type of increase often occurs when a transition takes place between states of similar type (e.g. $\pi \to \pi^*$ compared with $n \to \pi^*$).

When two chromophores in a molecule are adjacent (conjugated) it is generally found that the energy needed for absorption *decreases* (i.e. λ_{max} increases) and the extent of absorption (ε_{max}) *increases* compared with the values for separate groupings. This is illustrated for the diene and the unsaturated ketone listed in Table 4.1. The effect is particularly marked for benzene (and other aromatic compounds) which have extended π-systems. These molecules can also be distinguished since there are several $\pi \to \pi^*$ absorptions (because of the existence of several π and π^* orbitals).

When a benzene ring and another chromophore are conjugated then the characteristic absorptions are shifted to even longer wavelengths: an example is provided by Figure 4.6 which is the electronic absorption spectrum of phenylethanone, $C_6H_5COCH_3$ (there is also a weak absorption at higher wavelength; the details are given in Table 4.1).

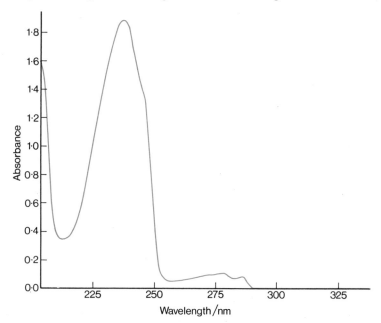

4.6   *Electronic absorption spectrum of phenylethanone,* $C_6H_5COCH_3$ *(recorded for a solution in hexane using a 1 cm cell; see Problem 4.3)*

Substitution of alkyl groups on aromatic and alkenic chromophore leads to small increases in λ_{max}.

The increased wavelength of absorption for *conjugated* molecules (i.e. when chromophores are adjacent to each other) can lead, if enough chromophores are present, to an absorption in the visible region (i.e. the compound is coloured). This occurs, for example, for aromatic compounds containing fused rings, for some 1,2-diketones (which are yellow), for the large molecules (with delocalised π-electrons) which are used as pH indicators (the extent of conjugation and hence the colour depends upon the ionisation of various groups in the molecule) and for molecules containing chains of conjugated double bonds.

For example, β-carotene (which occurs in carrots) is orange and has has $\lambda_{max} = 450$ nm, $\varepsilon_{max} = 15,000$ m^2 mol^{-1}.

β-carotene

When ε_{max} is very high, only very dilute solutions are needed for detection of the absorbing molecules, and the technique becomes a very sensitive method indeed for the detection of absorbing species (for example, absorption could be detected for a solution made up from less than 0·01 mg of β-carotene).

4.4 SOME APPLICATIONS OF U.V. AND VISIBLE ABSORPTION SPECTROSCOPY

(a) Structural Analysis

One important use of electronic absorption spectroscopy is the recognition of chromophores or groups of chromophores in organic molecules, by the measurement of λ_{max} and ε_{max} for the various absorption peaks. This information usually allows the type of molecule to be determined and, particularly when used in conjunction with other spectra, often provides valuable assistance with the determination of the exact molecular structure.

This branch of spectroscopy has been particularly useful in the structural investigation of steroids, which are biologically important molecules. An example is the hormone testosterone:

The absorption at 241 nm ($\varepsilon = 1600$ m^2 mol^{-1}) is characteristic of adjacent C=C and C=O double bonds in this sort of cyclic structure;

it has proved possible in examples like this to use the u.v. data to give an indication of the detailed structure around the chromophores.

(b) Quantitative Analysis

This particular application utilises the proven direct proportionality (see page 75) between the absorbance of a compound and its concentration. The relationship is widely applicable. There are many examples where, if ε for a given compound at a given wavelength is known, the measured absorbance of a solution of the compound at that wavelength leads to the *concentration* of the species concerned. The advantages of this method are that very low concentrations can be reliably obtained and that the *rate* of change of the absorption at a given wavelength can easily be monitored (this may be useful if the compound is involved in a chemical reaction). Some typical applications are as follows (other examples are given as references at the end of the chapter).

(i) *Kinetic Investigations.* Most spectrometers produce a plot of absorbance against λ but also allow the absorbance at one particular wavelength to be plotted as a function of time. This facility leads to a plot of concentration against time for either the loss of a reactant or the build-up of a product, as long as the absorption is characteristic of only the single component under investigation. Thus, the spectrometer not only provides vital information about the nature of the product (or products) of a reaction, it may also allow the rate of the reaction to be followed. Both pieces of information can aid the elucidation of the reaction mechanism concerned.

(ii) *Keto-enol tautomerism.* Both the u.v. spectrum (Figure 4.7) and the n.m.r. spectrum (page 115) of pentane-2,4-dione (acetylacetone, $CH_3COCH_2COCH_3$) indicate that this molecule does not exist simply as the molecular formula suggests—i.e. with two keto groups. Thus the n.m.r. spectrum (page 115) of pentane-2,4-dione (acetylacetone, molecule present in equilibrium: these are the **keto** and **enol tautomers** of the molecule (the phenomenon is called **tautomerism**):

keto   enol

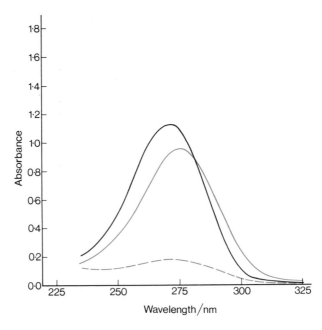

4.7   *Electronic absorption spectra of solutions (10^{-4} mol dm^{-3}) of pentane-2,4-dione,*
$CH_3COCH_2COCH_3$, in hexane ——, ethanol ——, and water -----; the cell has a
path length of 1 cm (see Problem 4.4).

This observation allows us to understand some unusual features of
the u.v. spectrum of pentane-2,4-dione: Figure 4.7 shows the absorption
for 10^{-4} mol dm^{-3} solutions in hexane, ethanol, and water. First, the
extent of absorption suggests that a simple carbonyl-type structure is
not responsible for the peak [for example, a much greater concentration
of propanone is necessary to obtain a strong ($n \rightarrow \pi^*$) absorption: see
Figure 4.2]. A $\pi \rightarrow \pi^*$ transition associated with a *conjugated* structure
is more likely. Second, the variation in the extent of absorption with
solvent is surprising (it certainly does not occur for a simple ketone).

The explanation is that the absorption at $\lambda = 270$ nm is due to the *enol*
form (cf. absorption for compounds of the type $C{=}C{-}C{=}O$, Table
4.1) and that the amount of enol form varies from solvent to solvent.
It can be demonstrated (e.g., by n.m.r. spectroscopy) that for a solution
in ethanol there is approximately 73% of the enol and 27% of the keto
form together in equilibrium (the percentage of the former accounts
for the absorption with $\lambda_{max} = 270$ nm, absorbance $= 0.96$ from the
10^{-4} mol dm^{-3} solution of the di-ketone in ethanol). The absorptions
for aqueous and hexane solutions imply that in water there must be a
much smaller proportion of enol than when ethanol is the solvent and
that in hexane solution there is correspondingly more enol.

The explanation is that in hexane the internally hydrogen-bonded enol form is preferred (the dotted line in the structure indicates a hydrogen bond) whereas in aqueous and alcoholic solutions the formation of *intermolecular* hydrogen bonds between the carbonyl group and the water molecules (or alcohol molecules) stabilises the keto form.

(c) Indicators

Figure 4.8 shows the variation in the absorption spectrum of a dilute aqueous solution of the indicator methyl red as the pH is altered. The indicator is red in acid (λ_{max} = 520 nm, see Figures 4.8 and 4.4) and yellow in alkali (λ_{max} = 425 nm) these being the colours of the acid and base forms:

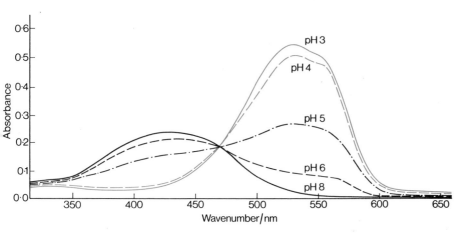

4.8 *Electronic absorption spectra of aqueous solutions of methyl red (all containing 0.004 g dm^{-3} of the indicator) at various pH values.*

At pH 1, the indicator is essentially all in the acid (HA) form; at pH 13, it is essentially all in the base (A) form. For intermediate pH values, both HA and A are present (to an extent which is governed by the pH and by the dissociation constant of the indicator, K_a). The heights of the "acid" peak (λ = 520 nm) and the "base" peak (λ = 425 nm) can be used, together with the 100% "acid" and "base" absorptions to determine the concentrations of HA and A at any given pH. This then leads to a measurement of K_a. Alternatively, if K_a is known, then measurement of [HA] and [A] gives a value for the pH. As will be appreciated, these measurements serve to quantify the procedure whereby, in a titration, the eye responds to the change in colour of a solution when a predominance of HA, say, is changed to a predominance of A.

(d) Metal ions and complexes

The application of u.v.-visible absorption spectroscopy also allows the determination of absorption maxima and molar decadic absorptivities for inorganic ions with electronic absorption in this part of the spectrum (e.g. for the absorptions which account for the purple colour of the manganate (VII) ion (permanganate, MnO_4^-) and for the yellow colour of the dichromate(VI) ion ($Cr_2O_7^{2-}$). This type of detailed information can then be useful in at least three contexts. First, there is a useful application in qualitative analysis; for example, the ions MnO_4^- and $Cr_2O_7^{2-}$ could be identified as present in a solution from the appearance of their characteristic absorption spectra. Second, there is the application for quantitative analysis: for instance, the inorganic anions mentioned above could be formed by the oxidation of solutions containing trace quantities of, say, Mn(II) and Cr(III), and the extent of absorption (the absorbance) at the appropriate wavelengths could be used to calculate the concentrations of the ions present (this can often be done for mixture of ions from a single spectrum). Lastly, there is considerable interest in the detailed analysis of absorptions from metal ions and complexes in terms of the electronic structure (and hence the possible electronic transitions) of the molecules or ions concerned. Particularly useful information can be obtained about transitions and energy levels involving d-electrons in transition metal ions.

4.5 PROBLEMS

4.1   Figure 4.2 is the u.v. spectrum of a solution of propanone in hexane. Calculate the molar decadic absorptivity (in $m^2 \, mol^{-1}$) for the peak with $\lambda = 279$ nm and compare your result with the value reported in the chapter (page 78).

4.2   The u.v. spectrum in Figure 4.3 was recorded for a solution of benzene in hexane. Calculate the molar decadic absorptivity (in $m^2 \, mol^{-1}$) for the peak with $\lambda = 255$ nm.

4.3   For a solution of phenylethanone ($C_6H_5COCH_3$) in hexane the molar decadic absorptivity for the peak with $\lambda_{max} = 279$ nm is 100 $m^2 \, mol^{-1}$. Calculate the concentration of the solution of phenylethanone in hexane whose u.v. spectrum is shown in Figure 4.6.

4.4 (a) Calculate the apparent molar decadic absorptivity for λ_{max} for pentane-2,4-dione in ethanol from Figure 4.7, which shows the spectrum for a solution containing 10^{-2} g dm^{-3} of the di-ketone. (b) The proportion of the enol form for pentane-2,4-dione in solution in ethanol has been estimated as 73%. From Figure 4.7 calculate the proportion of enol for solutions in (i) hexane, (ii) water.

4.5 Explain why aniline (phenylamine, $C_6H_5NH_2$) shows absorption maxima at approximately 230 and 280 nm (ε = 860 and 143 m^2 mol^{-1}, respectively) whereas salts of the anilinium (phenyl-ammonium) cation ($C_6H_5NH_3{}^+$) have absorptions at *ca.* 200 and 250 nm (ε = 750 and 16 m^2 mol^{-1}, respectively).

4.6 From the data given in Figure 4.8, estimate the dissociation constant K_a of the indicator, methyl red.

Further reading

I. F. Roberts, "Absorption Spectra of *d*-block elements", *Education in Chemistry,* 1971, **8**, 178.

W. O. George and C. H. J. Wells, "Visible absorption spectrum of iodine vapour", *Education in Chemistry,* 1972, **9**, 19.

E. F. Curragh and D. J. Thompson, "Chemical Kinetics—a spectro-photometric experiment", *Education in Chemistry,* 1973, **10**, 17.

Suitable Colorimetric Experiments

"Chemistry: Students' Book II," *Nuffield Advanced Science,* Pen-guin, London, 1970.

(a) Use of a colorimeter to follow the change in concentration of $MnO_4{}^-$ and hence to investigate the kinetics of the reaction between permanganate (manganate(VII)) ions and oxalate (ethanedioate) ions: page 58.

(b) An investigation of the rate of the reaction between iodine and acetone (propanone) in aqueous solution: page 46.

(c) Investigation of the stoichiometry of the Ni(II)-ethylenedi-amine-tetra-acetate (edta) complex ion: page 138

Advanced reading

Determination of the structure of organic molecules: see the books referred to in **Advanced reading** (structure determination) at the end of Chapter 3, page 69.

Chapter 5

Nuclear Magnetic Resonance Spectroscopy

Of all the techniques described in this book nuclear magnetic resonance (n.m.r.) spectroscopy was the last to be discovered but it has developed rapidly since the first n.m.r. experiment was successfully demonstrated in 1946. N.m.r. and mass spectroscopy now have no real rivals as the techniques most widely applicable for the solution of structural problems and their use has become, almost universally, a routine matter.

5.1 THE N.M.R. EXPERIMENT

An atom consists of negatively charged electrons surrounding a positively charged nucleus which itself is composed of protons (positively charged) and neutrons. It is found that nuclei which contain an odd number of protons or an odd number of neutrons (or both) have an extra property, in addition to their charge, which can be demonstrated by a suitable experiment. They can be shown to have a **magnetic moment** which means that they behave like tiny bar magnets, and the phenomenon can be demonstrated by detecting the energy of interaction when they are placed between the pole-pieces of a magnet. As happens when a simple bar magnet is placed in the magnetic field from a second magnet, the distinction between attractive and repulsive interaction can be detected. We will here be mainly concerned with the application of this technique for the study of hydrogen atoms in chemical compounds.

The hydrogen atom, ^1H, has a single proton for its nucleus and hence has a magnetic moment. In an applied magnetic field it experiences an interaction; the restriction applied by the quantum theory is that the magnetic moment can only be aligned either parallel to or opposed to the direction of the applied field.

It may be helpful to remember that a magnetic moment (or field) is associated with a body which is charged and in motion (cf. the magnetic field from electrons moving in a wire). This implies that the nucleus must be spinning in one of the two senses which give rise to the two alignments of its resultant magnetic moment.

The two possible alignments of the magnetic moment of the hydrogen nucleus are represented diagrammatically in Figure 5.1. These arrangements have different *energies,* and an exact amount of energy is

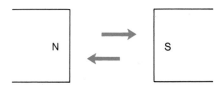

5.1   The two allowed alignments of the nuclear magnetic moment of a hydrogen atom in an applied magnetic field.

necessary to twist the magnet from one position to the other, i.e. from the attractive to the repulsive situation. Now for a magnetic moment μ the component in the direction of the applied field is $+\mu$ or $-\mu$, depending on the orientation. The energies of these two arrangements of the magnet are $-\mu B_0$ (aligned) and $+\mu B_0$ (opposed) respectively, where B_0 is the magnetic flux density of the applied field. The energy difference between the two arrangements is $2\mu B_0$ and this amount of energy (as radiation of the corresponding frequency, see page 38) is necessary to invert the proton's magnetic moment from the position of lower energy to that of higher energy. The exact condition is then:

$$\Delta E = h\nu = 2\mu B_0 \qquad (5.1)$$

This equation relates the magnetic moment of the proton, the magnetic flux density (i.e. the applied field strength), and the frequency which has to be employed before energy can be absorbed by the proton. The value of ν necessary to satisfy equation 5.1 clearly depends on the magnitude of the applied field. For values of B_0 typical of n.m.r. experiments, ν is in the *radio-frequency* region of the electromagnetic spectrum: ΔE is much smaller than the energy differences associated with electronic, vibrational, and even rotational changes.

It is usual to keep the applied frequency ν constant and to vary the magnetic field in a search for the exact condition when absorption of energy leads to "flipping" of the hydrogen atom's magnetic moment. In practice, a radio-frequency oscillator is the source of electromagnetic radiation of constant frequency ν and although a large permanent magnet is normally employed to generate the field, B_0 can be varied by means of a current supplied to coils wrapped around the pole-pieces (see Figure 5.2). When the radio-frequency radiation is absorbed (at **resonance**), an imbalance is produced in a radio-frequency bridge (which operates like a Wheatstone's bridge), and the resulting signal can be amplified and fed to a recorder as a plot of absorption of energy against magnetic field.

The choice of a fixed radio-frequency and large variable field is dictated by the need for as sharp an absorption as possible, something which will be achieved if the oscillator is highly stable and the field is

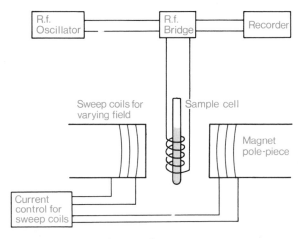

5.2 Basic features of an n.m.r. spectrometer

homogeneous and accurately adjustable. Despite the difficulty involved in producing high fields (and a correspondingly high radio-frequency), a considerable increase in sensitivity is achieved if this is done, the reason for which becomes clear when the dynamic nature of resonance is considered.

In a typical hydrogen-containing sample (e.g. an organic compound) there will be a large number of ^{1}H nuclei, distributed between the two energy levels; the ratio of the number in the upper energy level (N_u) to the number in the lower energy level (N_l) is given by the relationship known as the **Boltzmann Distribution.**

$$\frac{N_u}{N_l} = e^{-\Delta E/kT} = e^{-2\mu B_0/kT} \tag{5.2}$$

Here, T is the Kelvin temperature and k is Boltzmann's constant.

This relationship is generally applicable to the statistical distribution of particles between possible energy levels, and it tells us, for example, that for an energy difference (ΔE) which is *large* compared with the typical thermal energy of the molecules ($\sim kT$), the lower energy level is much more highly populated. However, for hydrogen atoms at room temperature and in the typical magnetic fields employed ΔE is considerably *smaller* than kT; the ratio N_u/N_l is nearly unity and the population of hydrogen atoms is almost equally divided between the two energy levels. For a million hydrogen atoms, there will be just a few more in the lower level than in the upper level. When irradiation of the sample with radiation at the resonance frequency (ν) takes place, transitions in both upward and downward directions occur (that is, absorption and emission take place). Overall *absorption* results because of the slight excess of nuclei in the lower level. The loss of energy from nuclei in the upper state to their surroundings ensures that the excess in the lower level is maintained and that the signal can be detected.

The smaller the ratio N_u/N_l, the greater is the sensitivity (i.e. the greater the excess in the lower level); as will be seen from equation 5.2, this means that a higher applied field (and higher ν, therefore) provides an advantage in sensitivity.

Many modern spectrometers operate with $\nu = 60$ MHz, for which the magnetic flux density (B_0) required for resonance of ^{1}H nuclei is 1.4 T.

In practice, signals can either be detected for the hydrogen atoms in a small quantity (*ca.* 0.5 cm^3) of a suitable liquid (e.g. ethanol), or, very usefully, for hydrogen atoms in solute molecules dissolved in a solvent (e.g. CCl_4) which may itself have no hydrogen atoms and therefore no absorption. Signals can be detected from substrates at low concentration (*ca.* millimolar) which means that n.m.r. spectra can be recorded even when only a few milligrams of a compound is available.

5.2   N.M.R. SPECTRA OF ORGANIC MOLECULES

On the basis of the previous discussion it would be anticipated that, since n.m.r. is a nuclear phenomenon, the resonance condition for different hydrogen atoms in a variety of molecules should not be affected by the electronic environment of each. To a certain extent this appears to be true; that is, for a fixed frequency (ν), hydrogen atoms in a variety of molecules absorb energy at approximately the same value of B_0. However, minor differences of crucial importance do exist.

For example, when the field is varied during an n.m.r. investigation of ethanol, CH_3CH_2OH, it is found that three distinct absorptions occur at slightly different field strengths (the differences between the fields for the absorptions are much smaller than the magnitude of the field). The n.m.r. spectrum is shown in Figure 5.3. The areas under the peak are approximately in the ratio 1 : 2 : 3, and we can conclude that the OH hydrogen atom, the two CH_2 hydrogen atoms and the three CH_3 hydrogen atoms are separately in resonance at different fields.

It should be noted here that no absorptions are seen for carbon and oxygen atoms since ^{12}C and ^{16}O do not have nuclear magnetic moments.

Similar observations are made for other organic molecules. For example, the spectrum of 2-methylpropan-2-ol (t-butyl alcohol, $(CH_3)_3COH$), shows two peaks in the ratio 9:1, and diethyl ether, $CH_3CH_2OCH_2CH_3$, exhibits a spectrum with two peaks in the ratio 3 : 2. In this case, the two methyl groups are themselves in an equivalent environment, but this is different from the environment of the two methene groups.

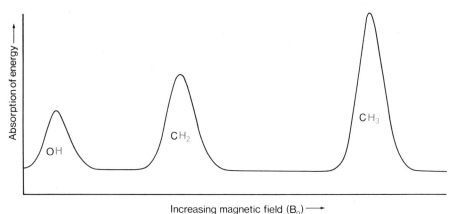

5.3   Low resolution n.m.r. spectrum of ethanol, CH_3CH_2OH

Figure 5.4 shows the two-peak spectrum from ethanal (acetaldehyde, CH_3CHO), and illustrates the **integration trace** (upper curve). The height of each step in this trace is proportional to the area under the appropriate absorption and hence to the number of hydrogen atoms in each group. For this example, the ratio is 1 : 3, corresponding to the separate absorptions of CHO and CH_3 hydrogen atoms, respectively.

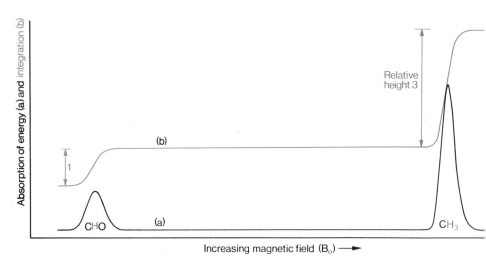

5.4   Low resolution n.m.r. spectrum of ethanal, CH_3CHO, showing absorption trace and integration trace.

Hydrogen atoms which undergo absorption of energy at different fields are said to have different **chemical shifts**, which depend on the environments of particular atoms in a molecule.

(a) Chemical shifts

The observation that in CH_3CH_2OH, for example, the three types of hydrogen atom absorb at slightly different applied magnetic fields suggests that somehow each type of hydrogen nucleus does not experience exactly the same magnetic field B_0. In practice, it can be shown that in the presence of the applied field small *local* magnetic fields are induced in the neighbourhood of the nuclei. Each nucleus now experiences an effective field,

$$B_{\text{effective}} = B_0 \pm B_{\text{local}}$$

and, since B_{local} is proportional to B_0, we can write

$$B_{\text{effective}} = B_0(1 \pm \sigma).$$

Then, depending on the size and direction of σ (a measure of the chemical shift), the applied field needs to be varied somewhat to achieve the unique *overall* field ($B_{\text{effective}}$) for the resonance condition ($h\nu = 2\mu B_{\text{effective}}$) to be obeyed.

It is now known that the local fields arise from electron circulations induced by the applied field, and two rather different cases can be distinguished.

(i) Figure 5.5 indicates the direction of circulation round a nucleus induced for a spherical electron cloud (this corresponds effectively to an electron in the 1s-orbital on the hydrogen atom). The electron movement creates an induced magnetic field which is opposed to the main field. The nucleus therefore experiences a smaller overall field than that applied (the nucleus is said to be **shielded**) and the resonance condition can now only be achieved for a higher-than-normal applied field.

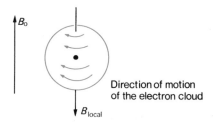

5.5   The production of local fields by induced electron-circulation

For CH_3CH_2OH the extent of shielding differs for each type of hydrogen atom because of different *electron densities* around the CH_3, CH_2, and OH hydrogen atoms. For the OH hydrogen atom the electron density around the nucleus will be relatively low because of the adjacent electronegative oxygen atom. However, for the CH_2 and CH_3 hydrogen atoms, which are progressively further from the oxygen atom, the electron density will be progressively greater. The CH_3 hydrogen atoms, being most highly shielded, absorb at highest field (Figure 5.3).

In general, we can relate the magnetic fields necessary for resonance of different hydrogen atoms to the structure of a molecule and, in particular, to the electron-withdrawing property (**negative inductive effect**) of the atoms present. More examples of this type of behaviour will be discussed later (page 94).

(ii) The second type of electron circulation which contributes to local fields in appropriate cases is that which can be induced in molecules containing double-bonds (i.e. molecules containing electrons in π-orbitals). A particularly clear example is provided by the marked effect when delocalisation around an aromatic ring is possible. This is called a **ring current** and is illustrated for benzene in Figure 5.6.

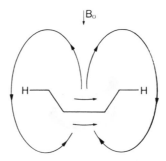

5.6   Induced electron circulation in benzene : the ring-current effect

When the benzene molecule is oriented perpendicularly to the applied field, circulation is induced in the overlapping $p(\pi)$-orbitals and this again leads to the production of local fields around the molecule. As the lines of force demonstrate, the local field augments the applied field at the hydrogen atoms, so that a *lower* field than expected needs to be applied for resonance to be achieved. The hydrogens are said to be **deshielded**. The rapid motion of the molecules means that only for a fraction of the time is the molecule in the particular orientation depicted but nevertheless the ring-current contribution is a dominant effect. As will be seen later, the existence of a ring current provides a readily-applied criterion for aromatic molecules.

This type of effect also contributes to the observed chemical shifts for hydrogen atoms close to carbonyl, alkynic, and certain other groups.

(b) The Measurement of Chemical Shifts

Measurements of chemical shifts are not quoted in field units, because, as we have seen, the field difference (ΔB) between two peaks depends upon the applied field (B_0) of the spectrometer employed.

The following procedure provides a way round this problem and leads to an acceptable universal scale. First, a suitable reference compound is chosen; tetramethylsilane (TMS), $Si(CH_3)_4$ is widely used as it is inert and it has a spectrum with a single absorption (there is only one type of hydrogen atom in the molecule). Then, the n.m.r. spectrum of the compound under investigation, with TMS added, is recorded. There will be a field difference ΔB between the absorption of a hydrogen nucleus in the sample and that for TMS. Then $\Delta B/B_0$ gives a measure of the chemical shift which is *independent of the applied field*. The resulting numbers are very small indeed and it is more convenient to multiply them by 10^6 and to refer to the resulting measure of chemical shift, δ, in parts per million (p.p.m.).

$$\delta/\text{p.p.m} = \frac{\Delta B}{B_0} \times 10^6$$

This leads to a scale of chemical shifts with $\delta = 0$ for the TMS hydrogens atoms and with most other hydrogen atoms in organic molecules in the range 0–10 δ. However, a closely related system is usually used and will be adopted for this book. This employs τ as a measure of chemical shifts, where τ is $(10 - \delta)$. On this scale the TMS hydrogen atoms have a τ value of 10, with hydrogen atoms in most other molecules absorbing in the 0–10 τ range (Figure 5.7). For example, the chemical shift of the hydrogen atoms in benzene is 2.75 τ, and the shifts of hydroxyl, methene and methyl hydrogen atoms in ethanol are 4.80, 6.35 and 8.80 τ respectively.

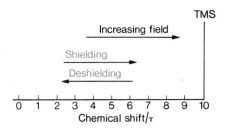

5.7   The τ-scale of chemical shifts

Chemical shifts, expressed as τ-values, are characteristic of the hydrogen nuclei in the compounds concerned and are independent of the spectrometer. Further, it is quite rare to find any marked dependence of the τ-value on temperature. The τ-value is usually unaffected if the solute is dissolved in an inert solvent.

(c) Relationship between Chemical Shifts and Molecular Structure

The dependence of τ upon molecular structure is understandable in terms of the magnitudes of the local fields produced by the electron-circulation effects described earlier. For example, although a CH_3 group in an alkane has a characteristic τ value of approximately 9.1, substitution in the molecule of a group with a negative inductive effect lowers the electron density round the hydrogen atom and hence the τ-value. This lowering is especially marked when the substituent is highly electronegative. Thus, along the series of halogenoalkanes CH_3X (X = I, Br, Cl, and F) the chemical shift of the methyl group hydrogen is progressively lowered (see Table 5.1); the lowest τ-value for the hydrogen atoms in fluoromethane indicates that the hydrogen atoms in this molecule are the least shielded (as expected, because fluorine is the most electron-attracting substituent).

TABLE 5.1

Chemical Shifts/τ of Hydrogen Atoms in Halogenomethanes

CH_3F	5.75
CH_3Cl	6.95
CH_3Br	7.30
CH_3I	7.85

The effect becomes even more pronounced when more than one electronegative element is present; for the hydrogen atom in $CHCl_3$, for example, $\tau = 2.75$.

Since the chemical shift of a hydrogen atom depends on its immediate environment within a molecule the resonance positions of hydrogen atoms in compounds of different structural types prove diagnostically useful; some approximate values for typical groups are given in Table 5.2.

The decrease in the τ-values along the series CH_3-C, CH_3-N, CH_3-O, [in alkanes, amines and methoxy compounds (methyl ethers), respectively] again reflects the increasing negative inductive effect of the substituent group.

Methyl esters of alkanoic acids, RCO_2CH_3, as might be expected, resemble methoxy compounds (methyl ethers) and have τ (CH_3) in the range 6.2–6.4. In contrast, both methyl ketones and esters of ethanoic acid (acetic acid), CH_3CO_2R, generally show a methyl group absorption at τ 8.0, this being the characteristic chemical shift of a methyl group adjacent to a carbonyl group.

Groups which have an inductive effect but which are further removed in the molecule have a decreased but sometimes noticeable effect; an example is the τ-value of 8.80 for the methyl-group hydrogen atoms in ethanol (lowered from *ca.* 9.1 by the effect of oxygen, but not as low as the τ-value of *ca.* 6.7 for molecules in which a CH_3 group is directly attached to the oxygen atom).

TABLE 5.2
Typical Values of Chemical Shifts/τ for Hydrogen Atoms in Organic Compounds

Group	Type of Compound	Chemical Shift/τ *
CH_3—C	alkane	9.1
C—CH_2—C	alkane	8.7
CH_3—C=C	alkene	8.4
CH_3—C(=O)O—	ester, acid	8.0
CH_3—C(=O)	ketone	7.9
CH_3—Ar	arene	7.8
HC≡C—	alkyne	8.0†
CH_3—N	amine	7.7
CH_3—O—	methyl ether	6.7
CH_3—O—C(=O)	methyl ester	6.3
CH_2=C	alkene	5.3†
H—C	arene	2.7†
H—C(=O)	alkanal	0.3†

* Typically ±0.1τ. Substituent effects may sometimes cause greater variation, especially for compounds indicated (†).

Methene hydrogen atoms (CH_2) have slightly lower τ-values than similarly placed CH_3 hydrogen atoms; the characteristic τ-value for a $-CH_2-$ group in an alkyl chain is 8.7.

A hydrogen atom attached to a carbon atom in the double bond of an alkene has a lower τ-value (typically about 5.3) than hydrogen atoms in saturated analogues. This shift is associated with the sp^2, rather than the sp^3, hybridisation of the alkenic carbon to which the hydrogen atom is bonded. Since few other types of hydrogen atom absorb in this area of the n.m.r. spectrum absorptions at τ $ca.$ 5 have particular diagnostic value.

For benzene and other aromatic compounds the ring current effect (section 5.2a) explains the unusually low τ-values observed.

The higher τ-value for a hydrogen atom attached to an alkyne triple bond ($ca.$ 8.0) also results from electron circulation (in the triple bond, around the molecular axis) which in this molecule provides a *shielding* effect at the hydrogen atoms. The reader may like to work out why the effect is opposite in direction to that observed for the hydrogen atoms in benzene.

When an unknown compound is studied with the aid of n.m.r. spectroscopy, the positions of the absorptions (i.e. the τ-values) give a fairly clear indication of the local environment of each type of hydrogen atom in the molecule, and the data given in Table 5.2 therefore prove diagnostically useful. In addition, remember that the integrator trace gives the extra information of the relative numbers of hydrogen atoms in each group.

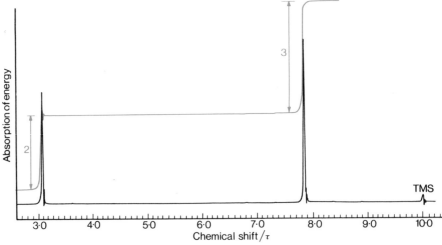

5.8 N.m.r. spectrum of 1,4-dimethylbenzene, 4-$CH_3C_6H_4CH_3$

The n.m.r. spectrum of 1,4-dimethylbenzene(*p*-xylene), Figure 5.8, provides a good example of the way in which a structural analysis can be carried out. Thus, in addition to the peak at 10τ from the added standard (TMS) two absorptions are detected: the ratio of the areas under the two absorptions is 2 : 3, and the peak at low-field (τ 3.05) is good evidence for the presence of aromatic-ring hydrogen atoms. The other peak, at higher field, has a τ-value of 7.85, characteristic of methyl groups attached to an arene. In this way the structure is confirmed. Note that there are only *two* different types of hydrogen atom in the molecule and therefore two absorptions.

A more complicated situation may be encountered for aromatic compounds with only one substituent or with several different substituents, since the ring hydrogen atoms are not all then in identical environments (unlike benzene or 1,4-dimethylbenzene). The non-equivalent hydrogen atoms may have slightly different chemical shifts, leading to a more complex pattern.

Although the information about the number and environment of a given type of hydrogen atom in a molecule is undoubtedly of considerable assistance in structure determination, there is further helpful information which can be obtained from many n.m.r. spectra if these are recorded under conditions of **high resolution**, and this is described in the next section.

(d) Spin-spin splittings

When care is taken to ensure that the magnetic field is accurately controlled and homogeneous, high-resolution conditions are achieved; it is then found that some of the characteristic absorptions discussed previously are split into several components. For example, Figure 5.9

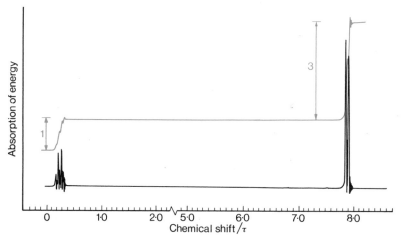

5.9   N.m.r. spectrum of ethanal, CH_3CHO

is the high resolution spectrum from ethanal which shows that the peak of area 3 (from CH_3, τ 7.85) is split into two (compare this with the low-resolution spectrum, Figure 5.4); the absorption is said to be a **doublet** and the two component lines have the same intensity (1 : 1). On the other hand, the absorption from the CHO hydrogen (τ 0.25) is split into a **quartet**, with relative intensities for the four lines of 1 : 3 : 3 : 1. The areas under the two *groups* of lines (i.e. the groups with different chemical shifts from the two types of hydrogen atom) are still in the ratio 3 : 1, as indicated by the integrator trace.

The splittings can be explained by considering first the methyl-group hydrogen atoms. These experience extra local magnetic fields which are due to the magnetic moment of the CHO hydrogen atom. This proton's magnetic moment must be aligned either *with* or *against* the applied field, so that in the CH_3—CH fragment the CH_3 hydrogen atoms experience an extra local magnetic field which can either *augment* or *oppose* the main field. Thus the overall field for resonance for the CH_3 hydrogen atoms can be achieved at two possible applied fields—one to be augmented by, the other to be opposed by, the local field from the aldehydic hydrogen atom. For a collection of CH_3CHO molecules some of the CHO-group hydrogen atoms will augment the main field whereas in other ethanal molecules the magnetic moment of the CHO hydrogen atom will be opposed to the main field: absorptions will be seen for the CH_3 hydrogen atoms in both of these possible environments and therefore there are two peaks for the resonance from the CH_3 hydrogen atoms. The CH_3 resonance is said to be "split" by the single hydrogen atom, and the distance between the two peaks of the doublet is called the **splitting** (or **spin-spin splitting**, since the magnetic moments are effectively produced by the spinning motion of the nuclei).

A splitting into a doublet is characteristic of a group of equivalent hydrogen atoms (CH_3 in this example) split by *one* neighbouring hydrogen atom. We can now derive the splitting patterns for larger numbers of neighbouring hydrogen atoms, as, for example, when the resonance from a hydrogen atom is split by an adjacent CH_2 or CH_3 group.

Splitting from two hydrogen atoms. Consider the spectrum from diethyl ether, $CH_3CH_2OCH_2CH_3$ (Figure 5.10) which has two main absorptions, from the hydrogen atoms of the CH_2 and CH_3 groups. The two sets of CH_3 hydrogen atoms (τ 8.85) will each experience local fields from adjacent —CH_2— hydrogens, and we must therefore work out the possible "arrangements" of the latter. Two CH_2 hydrogen nuclei may be both aligned in the applied field in one direction (which

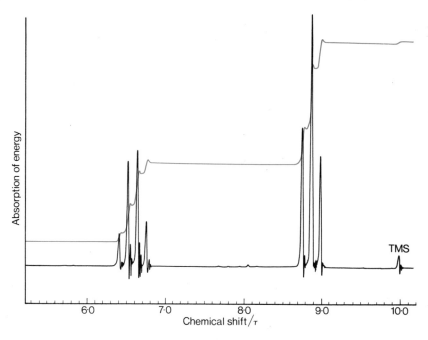

5.10   N.m.r. spectrum of diethyl ether, $CH_3CH_2OCH_2CH_3$

we can represent $\rightrightarrows$), both in the other direction ($\leftleftarrows$), or in opposite directions ($\rightleftarrows$ or $\leftrightarrows$). This gives three possible local magnetic fields, of which, for a collection of molecules, the arrangement with the magnets opposed ($\rightleftarrows$ and $\leftrightarrows$) can be achieved in two possible ways and will be twice as common as the others. Thus the splitting of the CH_3 resonance produces a 1 : 2 : 1 pattern, a **triplet**.

Splitting from three hydrogen atoms. In the two previous examples (CH_3CHO and $CH_3CH_2OCH_2CH_3$), we have to complete the understanding of the spectrum by realising that the CHO hydrogen atom in CH_3CHO must be split by the CH_3 group and that the two sets of CH_2 hydrogen atoms in $CH_3CH_2OCH_2CH_3$ (at τ 6.6) must be split by the adjacent CH_3 hydrogen atoms. We therefore have to work out the possible local magnetic fields provided by the three magnetic moments of a CH_3 group. These are as follows:

There are four possible resultant magnetic fields, two of which can be achieved in three possible ways: thus a CH_3 group splits the resonance of a neighbouring nucleus into a four-line pattern (**quartet**) with relative intensities $1:3:3:1$, as confirmed by the splitting of the CH_2 absorption in Figure 5.10 and of the CHO absorption in Figure 5.9.

More than three hydrogen atoms. In a fragment —CH_2—CH—CH_2—, in which the two CH_2 groups are equivalent (and have the same chemical shift), the single hydrogen atom will interact with four equivalent hydrogen atoms; the resulting pattern is a $1:4:6:4:1$ **quintet**. Five equivalent hydrogen atoms produce a splitting pattern of $1:5:10:10:5:1$, and six equivalent hydrogen atoms (e.g. in a $(CH_3)_2CH$—group) split the single hydrogen atom's resonance into a $1:6:15:20:15:6:1$ pattern. The splitting patterns follow the coefficients of the terms in the binomial expansion and are conveniently expressed in the form of "Pascal's Pyramid":

Number of hydrogen atoms causing splitting	Splitting pattern produced (relative intensities of lines)
1	1 1
2	1 2 1
3	1 3 3 1
4	1 4 6 4 1
5	1 5 10 10 5 1
6	1 6 15 20 15 6 1

(e) **Summary**

Before proceeding further, it is perhaps worthwhile reminding ourselves of the various stages in the analysis of a complex n.m.r. spectrum by reference to that of CH_3CHO:

First, under low resolution conditions there are two absorptions, of relative area $3:1$, from the CH_3 and the CHO hydrogens, respectively. These different groups have different chemical shifts (τ-values) which are typical of the chemical environment of the two types of hydrogen atom (Table 5.2).

Second, under high resolution conditions, splitting can be seen. Interaction of the CH_3 hydrogen atoms with the CHO hydrogen atom means that the former resonance is a $1:1$ doublet; interaction of the CHO hydrogen atom with the CH_3 hydrogen atoms means that the aldehydic hydrogen's resonance is a $1:3:3:1$ quartet.

The size of the splitting (Figure 5.9)—i.e. the separation between the peaks of each multiplet—is a measure of the energy of the interaction, and is the same for both resonances. The splitting is customarily quoted in frequency units ($\Delta\nu$/Hz); it is, for a given compound, independent of the magnitudes of the characteristic radio-frequency and applied magnetic field of the n.m.r. spectrometer employed.

A word of explanation is necessary here since the measurement involves (usually) conversion from the τ-scale: the problem is to express a separation (measured as $\Delta\tau$) as $\Delta\nu$ (Hz). From the definition of τ (page 93) we have:

$$\Delta\tau = \frac{\Delta B}{B_0} \times 10^6$$

and, from Equation 5.1,

$$\frac{\Delta B}{B_0} = \frac{\Delta\nu}{\nu}$$

so that

$$\Delta\nu = \frac{\nu . \Delta\tau}{10^6}$$

With a 60 MHz spectrometer (i.e. $\nu = 60 \times 10^6$ Hz), such as that used for all the n.m.r. spectra shown in this book, 1 τ unit is equivalent to 60 Hz. (For a 100 MHz spectrometer, 1 τ unit equals 100 Hz.)

In the example CH_3CHO, the separation between the split lines (for both the 1 : 1 and the 1 : 3 : 3 : 1 patterns) is 0.05 τ, as measured with this spectrometer. This splitting is 3Hz (it is referred to as J_{HH}) and it is always 3 Hz, no matter which spectrometer is employed. The separation *measured as* $\Delta\tau$ is somewhat less on a 100 MHz spectrometer.

Analysis of the splittings in the n.m.r. spectra of organic compounds is usually fairly straightforward because appreciable splittings normally occur only between hydrogen atoms on *neighbouring* atoms (information about the alignment of one magnetic moment is transmitted to other nuclei through the bonds and the effect dies off rapidly with the number of intervening bonds). This makes the technique extremely effective for distinguishing isomeric alkyl structures (straight-chain and branched): see section 5.3.

It must also be remembered that hydrogen atoms in exactly equivalent environments in a molecule do not split each other. Thus, for CH_3CHO, for example, the methyl hydrogen atoms, while split by the CHO hydrogen, do not split each other. This is because they are all identical and in resonance together (at the same chemical shift).

5.3 EXAMPLES OF SPECTRA SHOWING SPIN-SPIN SPLITTINGS

(a) Butanone (methyl ethyl ketone) $CH_3COCH_2CH_3$

The n.m.r. spectrum (Figure 5.11) contains a single peak from the methyl group adjacent to the carbonyl group, with the expected

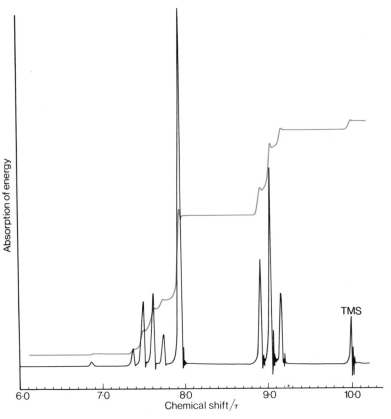

5.11   N.m.r. spectrum of butanone, $CH_3COCH_2CH_3$

chemical shift (τ 7.9; see Table 5.2). There is no splitting of this peak because there are no hydrogen atoms on the adjacent carbon atom. The other peaks are from the hydrogen atoms in the CH_2 group (τ 7.55) and the other CH_3 group (τ 9.0); the CH_2 peak has the lower τ-value because of the effect of the adjacent carbonyl group and the absorption appears as a 1:3:3:1 quartet because of the methene group's interaction with the CH_3 group. The CH_3 resonance is split into a 1:2:1 triplet by the CH_2 hydrogen atoms (these two multiplets form the characteristic pattern from an ethyl group).

In this example, as in many others, the splitting pattern only *approximates* to the expected 1:2:1 and 1:3:3:1 relative intensities: peaks within these groups are slightly larger in the direction of the resonance of the group responsible for the splitting. The distortion (which can be ignored here) becomes more pronounced when chemical-shift differences become small.

As can be seen from Figure 5.11, the integration trace still indicates the relative numbers of hydrogen atoms in the groups with different chemical shifts.

(b) 1,3-Dibromopropane (BrCH$_2$CH$_2$CH$_2$Br)

The n.m.r. spectrum, shown in Figure 5.12, consists of two groups of resonances, the numbers of hydrogen atoms concerned being 2 : 1 (i.e. the relative total areas, as indicated by the height of superimposed steps in the integrated curve for each resonance). These are evidently the four outside and the two central methene hydrogen atoms, respectively, with the expected τ-values; the outside methene groups absorb at lower τ-value because of the inductive effect of the bromine atoms.

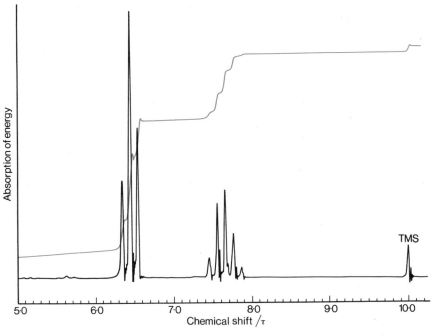

5.12   N.m.r. spectrum of 1,3-dibromopropane, BrCH$_2$CH$_2$CH$_2$Br

The central CH$_2$ resonance is split into a 1 : 4 : 6 : 4 : 1 quintet by the *four* neighbouring hydrogen atoms; the outside CH$_2$ groups each have one neighbouring CH$_2$ group and hence appear as 1 : 2 : 1 triplets.

(c) Bis-(l-methylethyl) ether (Di-isopropyl ether),
(CH$_3$)$_2$CHOCH(CH$_3$)$_2$

The spectrum (Figure 5.13) indicates the presence of two different types of hydrogen atom, the numbers of each type being in the ratio 1 : 6. The methyl group resonance (at τ 8.9) is split into a doublet, since each CH$_3$ group hydrogen atom experiences an interaction with the neighbouring single hydrogen atom. The CH resonance (τ 6.4) is split into a septet, characteristic of interaction with six equivalent hydrogen atoms (see page 100).

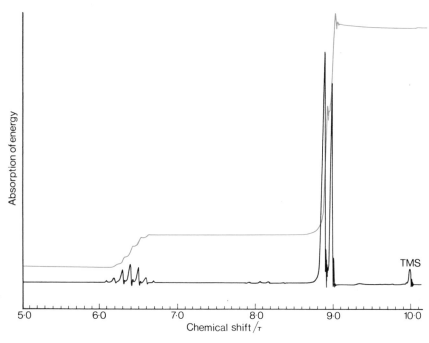

5.13   N.m.r. spectrum of bis(1-methylethyl) ether, $(CH_3)_2CHOCH(CH_3)_2$

(d) **Methyl-4-nitrobenzene (p-nitrotoluene), $4\text{-}CH_3C_6H_4NO_2$**

The spectrum (Figure 5.14), recorded for a solution in CCl_4 shows clearly the absorption from aromatic (τ 1.9–2.7) and aliphatic (τ 7.55)

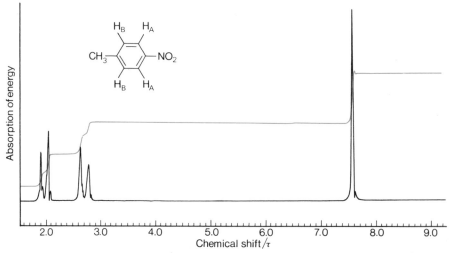

5.14   N.m.r. spectrum of methyl-4-nitrobenzene, $4\text{-}CH_3C_6H_4NO_2$

hydrogen atoms, in the ratio $4:3$. The methyl group resonance is not split and the aromatic hydrogens appear as two non-equivalent pairs; H_A, τ 1.9, and H_B, τ 2.7. The signals for H_A and H_B are both doublets because of the splitting for each hydrogen atom by the adjacent hydrogen atom (J 9 Hz). This four-line pattern is typical of a 1,4-disubstituted benzene ring with two different substituents.

Not only does the observation of the splitting patterns in n.m.r. spectra indicate which groups are attached to others in an organic molecule, but also, in certain cases, it is found that the magnitude of the splitting is informative. For example, for alkenes in which the hydrogen atoms are not all equivalent (if they were, they would all have the same τ-value and hence have no observable splitting) then the following splittings are typical.

$J_{HH} \simeq 17$ Hz $J_{HH} \simeq 10$ Hz $J_{HH} \simeq 0\text{-}2$ Hz

The recognition of these differences sometimes enables a choice to be made between various possible isomeric structures (see also the similar application of i.r. spectroscopy, page 64). For example, in one of the isomeric forms of 3-phenylpropenoic acid (cinnamic acid) the two alkenic protons have τ 2.17 and 3.54, with a splitting (J_{HH}) of 17 Hz: this must therefore be the *trans* isomer.

We can also use this knowledge to illustrate the exact analysis of the fairly complicated n.m.r. spectrum of phenylethene (styrene), Figure 5.15. All the alkenic hydrogens have separate τ-values [that from the nearest hydrogen to the aromatic ring (H_A) having the biggest shift away from the usual τ-value (about 5) for unsubstituted alkenes] and there is a large peak (τ 2.8) from the aromatic hydrogens. Each alkenic hydrogen atom will be split by the two non-equivalent alkenic hydrogen atoms: in each case the pattern which is produced has four lines (a doublet of doublets) with relative intensities ideally $1:1:1:1$ (but which show some deviation from the expected heights—see page 102). The separate splittings can be measured from the spectrum (J_{AB} 17.2, J_{AC} 10.6, J_{BC} 1.2 Hz) and can be assigned to pairs of hydrogen atoms on the basis of the guidelines indicated above.

5.4 THE N.M.R. SPECTRUM OF ETHANOL

Although n.m.r. spectra of organic molecules are clearly governed by the environment of the hydrogen atoms *within* the molecule (and hence do not usually depend on the solvent, for example) in a few instances the spectrum can give information about *intra*-molecular effects. This is often found for compounds in which there are exchangeable atoms, such as the hydroxylic hydrogen atom in alkanols.

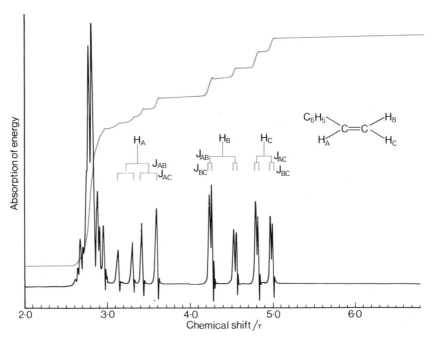

5.15 *N.m.r. spectrum of phenylethene,* $C_6H_5CH{=}CH_2$, *showing the assignment of the splittings*

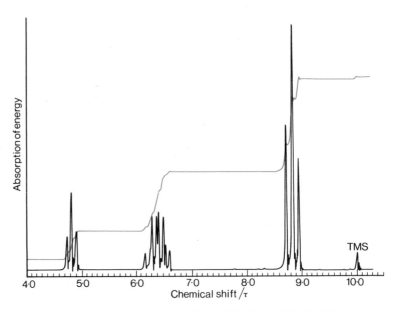

5.16 *N.m.r. spectrum of highly-purified ethanol,* CH_3CH_2OH

Figure 5.16 is the high resolution spectrum of pure ethanol; it consists of a $1:2:1$ triplet for the CH_3 group (split by the CH_2 hydrogen atoms, J 7 Hz), a $1:2:1$ triplet for the OH hydrogen atom (split also by the CH_2 hydrogen atoms, J 5 Hz) and a pattern for the CH_2 hydrogen atoms which is indicative of splitting by CH_3 (into a $1:3:3:1$ quartet) *and* OH $(1:1)$ which, since J is not the same for the two splittings, becomes approximately $1:1:3:3:3:3:1:1$ (i.e. all the lines in the $1:3:3:1$ pattern are doubled).

Figures 5.17 illustrates the effect of adding a little aqueous acid to the ethanol. The following acid-catalysed exchange of the hydroxyl-hydrogen atom takes place:

$$C_2H_5OH + HOH \rightleftharpoons C_2H_5OH + HOH$$

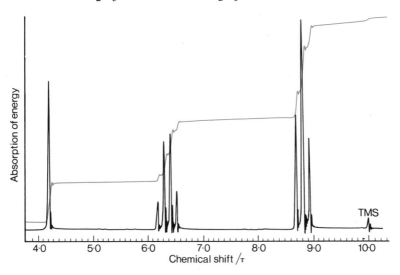

5.17   *N.m.r. spectrum of ethanol containing a trace of acid*

Thus the CH_2 hydrogen atoms no longer experience a simple local field from *one* single hydrogen atom (with its two possible orientations) but, over a finite period, experience the averaged effect of all the hydrogen atoms visiting the oxygen atom. The coupling to a single hydrogen atom is removed and the splitting of the OH and CH_2 peaks disappears.

This ready exchange often means that, unless the samples are rigorously purified, hydroxyl-hydrogen (OH) splittings are not observed. We can make use of this type of exchange reaction in order to prove that a compound has an exchangeable hydrogen atom (e.g. $-OH$). Addition of D_2O, followed by exchange, leads to the formation of ROD, instead of ROH, and since deuterium (^{2}H) does not absorb in the same part of the n.m.r. spectrum as ^{1}H (i.e. in the same region of field) the absorption from the hydroxyl-hydrogen atom is removed.

Compounds with hydroxyl groups also often have a τ-value for the OH hydrogen atom which is somewhat solvent-dependent. This is not only because the $-$OH group can be hydrogen-bonded in a suitable solvent (and hence the proton's environment altered; see, for example, the effect of H-bonding on the i.r. absorption for OH groups, page 60) but also because rapid exchange, if this is possible, causes the hydrogen atom to experience an *averaged* local environment. An example which illustrates the latter effect is provided by the n.m.r. spectrum of an aqueous solution of ethanoic acid (CH_3CO_2H in H_2O): there is only *one* OH peak, at a τ-value between those of the hydroxyl-hydrogen atoms in pure CH_3CO_2H and pure H_2O, because OH hydrogen atoms are *rapidly* exchanged between the two types of molecule. The chemical shift of the resulting single absorption depends on the mole-fraction of each component.

5.5 WORKED EXAMPLES

At this stage the reader is encouraged to attempt the following problems, for each of which the molecular formula is given. These are

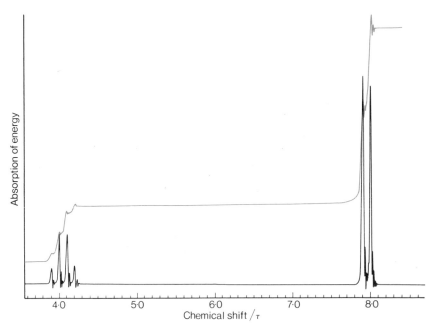

5.18   N.m.r. spectrum of Worked Example 5.1; molecular formula $C_2H_4Cl_2$

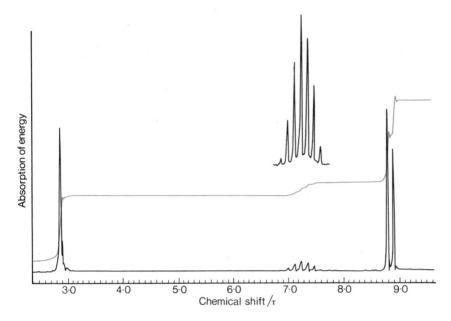

5.19   N.m.r. spectrum of Worked Example 5.2; molecular formula C_9H_{12}. Inset shows 7.25 τ peak recorded under conditions of increased sensitivity.

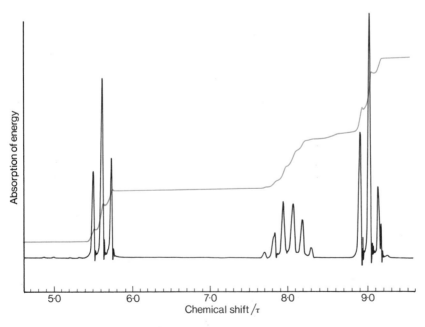

5.20   N.m.r. spectrum of Worked Example 5.3; molecular formula $C_3H_7NO_2$

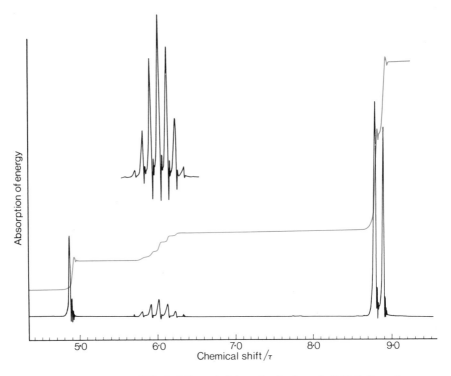

*5.21 N.m.r. spectrum of Worked Example 5.4; molecular formula C_3H_8O. **Inset** shows 6.0 τ peak recorded under conditions of increased sensitivity.*

followed by a discussion which gives the answers and some brief explanatory notes. Remember that the spectra offer three pieces of vital information; the chemical shifts (related to the types of hydrogen atom present), the integrated trace (the steps are proportional to the relative numbers of hydrogen atoms in each group) and the splittings (which depend on the number of hydrogen atoms in neighbouring groups).

Discussion

(i) Figure 5.18 is the spectrum of 1,1-dichloroethane, CH_3CHCl_2. There are clearly two types of hydrogen atom in the molecule, the number(s) in each group being in the ratio of the integrated intensities (3 : 1). This strongly suggests the presence of CH_3 and CH groups, which is confirmed by the splittings of 1 : 3 : 3 : 1 (of the CH absorption due to interaction with three equivalent hydrogen atoms) and 1 : 1 (the doublet from splitting of the methyl group absorption by the single hydrogen atom). The low τ-value (4.05) for the single hydrogen atom reflects the negative inductive effect of the chlorine atoms.

(ii) Figure 5.19 is the spectrum of an aromatic compound, as judged from the absorption at τ 2.85. The aliphatic part of the molecule has two types of hydrogen atom, apparently in the ratio 1 : 6 (as deduced from the integrated trace). The high-field peak is typical of a C—CH_3 group, so that the part-structure —CH $(CH_3)_2$ may be suggested. This is confirmed by the splittings—the methyl hydrogen absorption is split into a doublet by interaction with one hydrogen (CH) and the absorption of the single hydrogen is split by interaction with six equivalent hydrogen atoms to give a septet (1 : 6 : 15 : 20 : 15 : 6 : 1— see Figure inset). Since the ratio of the number of aromatic hydrogen atoms to aliphatic hydrogen atoms is approximately 5 : 7 the structure must be (1-methylethyl) benzene (isopropylbenzene, cumene):

(iii) From Figure 5.20 we can conclude that there is a propyl group $(CH_3CH_2CH_2-)$ present. Thus the high-field absorption at τ 9.05 (of relative intensity probably 3) is typical of a methyl group in an alkane, and there are two other absorptions with relative intensity 2: of these, the low-field (CH_2) peak is evidently split by a CH_2 group (to give a 1 : 2 : 1 triplet), and the splitting of the middle multiplet (evidently another CH_2) by CH_3 *and* CH_2 gives a 1 : 5 : 10 : 10 : 5 : 1 pattern. This is the spectrum of $CH_3CH_2CH_2NO_2$, although from the evidence so far presented the alternative structure $CH_3CH_2CH_2ONO$ cannot be ruled out. Distinction between these two would be made on the basis of the expected τ-values (e.g. from Tables of Data) for hydrogen atoms in alkyl nitrites ($-CH_2ONO$) and nitroalkanes ($-CH_2NO_2$), from other spectroscopic data (e.g. u.v., i.r.) or from chemical evidence.

(iv) Figure 5.21 is clearly the spectrum of a compound with a $-CH(CH_3)_2$ group (cf. Figure 5.19) as judged by the relative intensities (1 : 6) and the splittings of the peaks at τ 6.0 and 8.8, respectively. This leaves an oxygen atom and a hydrogen atom to be accounted for, so the compound must be propan-2-ol, $(CH_3)_2CHOH$. The chemical shift of the remaining single hydrogen atom is at least consistent with this formulation; the lack of splitting of OH and CH can be understood on the basis of the exchange which takes place in alkanols which are not rigorously purified (cf. ethanol, page 107).

Figures 5.18–21 clearly demonstrate the effectiveness of n.m.r. spectroscopy in distinguishing possible alkyl groups, e.g. $-CH_2CH_2CH_3$ and $-CH(CH_3)_2$.

5.6   PROBLEMS

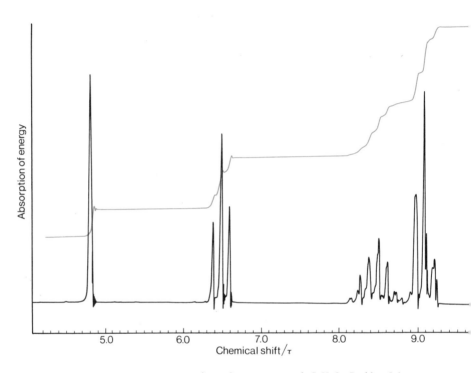

5.22   *N.m.r. spectrum of an unknown compound, C_3H_8O : Problem 5.1*

5.1   Identify the compound, of molecular formula C_3H_8O, whose n.m.r. spectrum is shown in Figure 5.22.

5.2   Identify the compound, of molecular formula $C_4H_8O_2$, whose n.m.r. spectrum is shown in Figure 5.23.

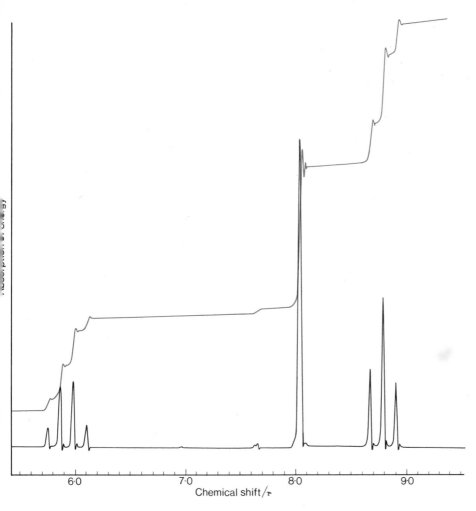

5.23   *N.m.r. spectrum of an unknown compound, $C_4H_8O_2$: Problem 5.2*

5.3   The n.m.r. spectrum shown in Figure 5.24 is for the compound whose mass spectrum was given in Problem 1.3 (page 34). Does this confirm your assignment?

5.4   In the n.m.r. spectrum of pure ethanol (Figure 5.16), recorded on a 60 MHz spectrometer, the separation between the 1 : 2 : 1 peaks of the OH hydrogen splitting is 0.08 τ. Calculate the splitting, Δv, (in Hz) between the hydroxyl and methene hydrogen atoms in the molecule.

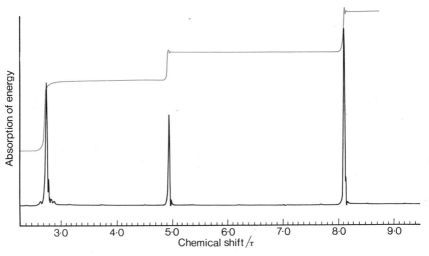

5.24   N.m.r. spectrum of an unknown compound: Problem 5.3

5.7   FURTHER APPLICATIONS OF N.M.R.

(a) Hydrogen-bonding

For a solution of phenol (C_6H_5OH) in tetrachloromethane (carbon tetrachloride), the chemical shift (τ) of the hydroxyl hydrogen atom's absorption depends markedly on the concentration. We can interpret this observation in terms of the environment of the hydrogen atom concerned; at low concentrations of phenol, each molecule will be surrounded by CCl_4 molecules which will not interact appreciably with the hydroxyl group, but at higher concentrations of phenol there occurs intermolecular association of these molecules *via* hydrogen bonding. This involves an attraction between a phenol molecule's oxygen atom (which is electronegative) and the positively polarised hydrogen atom in the hydroxyl group of another phenol molecule. By contrast, if an *intra*-molecular hydrogen bond is possible, then the −OH hydrogen atom is much less susceptible to changes in its external environment brought about by changing the concentration. This is true, for example, for 2-nitrophenol (1), where an internal hydrogen bond is formed between the hydrogen atom on the phenolic oxygen atom and one of the electronegative oxygen atoms of the nitro-group. For this molecule τ (OH) is much less sensitive to changes in concentration.

(1)

(b) Keto-enol tautomerism: demonstration and quantitative estimation

It is possible to use an n.m.r. spectrometer to determine relative amounts of several constituents in a mixture. An interesting example is provided by the spectrum (Figure 5.25) recorded for a sample of pure pentane-2,4-dione (acetylacetone, $CH_3COCH_2COCH_3$). The spectrum confirms that the compound exists in two forms, and careful analysis indicates that these are the keto- and enol-tautomers, (2) and (3), respectively (the phenomenon is known as tautomerism; see also page 81).

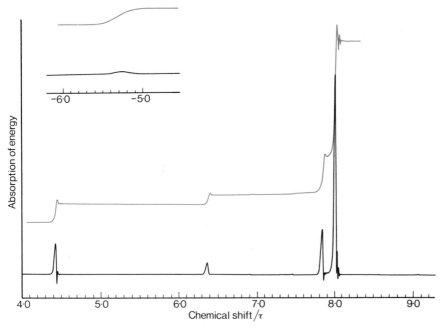

5.25 *N.m.r. spectrum of pentane-2,4-dione, $CH_3COCH_2COCH_3$*

The *enol* form (3) is recognised by the typical alkene hydrogen absorption at τ 4.45; the broad peak at -5.3 τ characterises an hydroxyl-hydrogen atom (the low τ-value reflects the effect of both oxygen-atoms: the hydrogen atom is hydrogen-bonded to the oxygen atom of carbonyl group), and it disappears on shaking the compound with D_2O. The peak at 6.4 τ is characteristic of the CH_2 group between two carbonyl groups in the *keto* form. Comparison of the integrations of the τ 4.45 and 6.40 peaks leads to a calculated ratio of 3.75:1 for the relative amounts of enol and keto tautomers (79 % enol). The dependence of the ratio of the tautomers on the nature of the solvent (as illustrated by the ultra-violet data given on page 82), and on the temperature, can readily be investigated.

This type of information is not readily determined without the use of spectroscopic techniques; for example, two tautomers are usually very easily interconvertible, so that it is not often possible to isolate and estimate the percentage of one of the individual forms. Similarly, attempts to remove one of the tautomers by a specific chemical reaction (e.g. the reaction the enol form (3) with bromine) may be thwarted by the rapid re-establishment of the equilibrium to produce the component being removed; this will then result in complete reaction of the compound *via* that particular tautomer.

(c) Dynamic effects

We have already seen how the addition of a trace of acid to a pure alcohol leads to an increased rate of hydroxyl-hydrogen atom exchange and hence to the collapse of the splitting between OH and the hydrogen atoms on the carbon atom adjacent to the oxygen atom in the alcohol. The observation of this type of *dynamic* effect in n.m.r. spectra is not limited to examples of rapid chemical reactions (like hydrogen–atom exchange) but can also occur for some rapid intra-molecular processes.

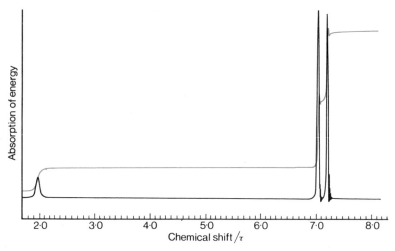

5.26   N.m.r. spectrum of N,N-dimethylmethanamide, $(CH_3)_2NCHO$, at room temperature

For example, consider the n.m.r. spectrum of N,N-dimethyl-methanamide (N,N-dimethylformamide), (4), at room temperature (Figure 5.26). As indicated by their different chemical shifts ($\Delta\tau$ 0.16, recorded on a 60 MHz spectrometer) the two methyl groups are not equivalent (i.e. not in identical environments). This arises from restriction of rotation of the $-N(CH_3)_2$ group about the C—N bond (which has partial double-bond character):

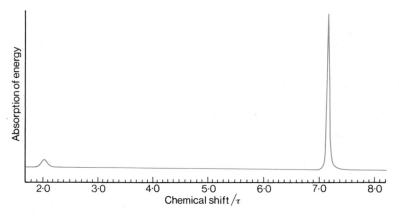

At higher temperatures, the molecules possess more energy, the rate of the rotation about the C—N bond increases, and the methyl groups eventually appear to be equivalent (that is, there is one large peak; Figure 5.27 is the spectrum at 150°).

Absorption of energy

2·0   3·0   4·0   5·0   6·0   7·0   8·0

Chemical shift /τ

5.27   *N.m.r. spectrum of N,N-dimethylmethanamide, $(CH_3)_2NCHO$, at 150°*

The critical rate of rotation, at which the two separate peaks just become coalesced to a single peak, occurs when the rate of rotation is approximately equal to the difference between the two separate absorptions expressed in frequency units (i.e. $\Delta v/Hz$).

In this case $\Delta\tau$ is 0.16, which must be converted into $\Delta v/Hz$. Since the spectrometer operates at 60 MHz, Δv is 0.16 × 60 (see page 101), i.e. about 10 Hz. Thus, when the lines coalesce, at approximately 100°, the rate of rotation is of the order of 10 Hz (i.e. 10 times per second). Above this temperature, and hence at faster rates, only a single *averaged* line results for the two methyl groups.

In this example, the two groups whose positions in the n.m.r. spectrum are being interconverted differ in their τ-values. Exactly the same arguments apply when a *splitting* (J_{HH}) disappears because of rapid exchange. Thus, in pure ethanol the CH_2 hydrogen atoms have a splitting of 5 Hz from the hydroxyl hydrogen atom (a 1 : 2 : 1 pattern; Figure 5.16, page 106). When acid is added, the hydroxyl hydrogen atom undergoes acid-catalysed exchange; as the exchange rate increases, the splitting will disappear and the coalescence point will correspond to an exchange rate of *ca*. 5 Hz (five times per second). At faster rates of exchange, a single line with no splitting is observed.

Detailed analysis of the shape of the signal in this and similar cases can be employed to measure the rate of exchange (or rotation) at different temperatures. Then, an Arrhenius plot leads to an estimate of the activation enthalpy (energy barrier) for the process concerned. For N,N-dimethylmethanamide, the barrier to rotation is estimated as 30 kJ mol^{-1}.

(d) N.m.r. from other nuclei

As well as ^{1}H, other nuclei have nuclear magnetic moments; these include ^{2}H (deuterium), ^{11}B, ^{13}C, ^{14}N, ^{19}F, ^{31}P, ^{35}Cl, ^{37}Cl, ^{79}Br and ^{81}Br.

These nuclei may give rise to splittings in the ^{1}H–n.m.r. spectrum of appropriate molecules. However these splittings are not always observable and, in particular, it is rarely possible to observe splittings from ^{2}H, ^{14}N or the chlorine and bromine isotopes.

However, it is possible to design n.m.r. spectrometers which operate at different combinations of ν and B than those used for ^{1}H resonance in order to detect resonance from the nuclei listed above. For example, absorption from ^{19}F atoms in fluorine-containing molecules can be detected for a spectrometer with $\nu = 60$ MHz at a value of B of 1.5 T. In this way, the scope of n.m.r. spectroscopy for probing molecular structure has been extended, as the following example illustrates.

The ^{11}B n.m.r. spectrum of diborane (B_2H_6) shows a single resonance, with a large 1 : 2 : 1 splitting (implying an interaction of the boron nuclei with two hydrogen atoms) and a further 1 : 2 : 1 splitting (implying a weaker interaction of the boron atoms with two more hydrogen atoms). We can conclude that the two boron atoms in B_2H_6 must have identical environments within the molecule (there is only one boron atom chemical shift), and must each interact with two different pairs of hydrogen atoms. Thus the n.m.r. spectrum provides evidence to support the assignment of the following symmetrical bridged-structure to the molecule (electron diffraction, described in Chapter 6, allows the bond lengths and angles to be accurately determined):

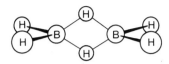

The ^1H n.m.r. spectrum of B_2H_6 shows two resonances in the ratio $2:1$, indicating, as expected, that there are two types of hydrogen (the end and the bridge atoms, respectively). In contrast, the ^1H n.m.r. spectrum of ethane (C_2H_6) shows just a single absorption because all the hydrogen atoms are equivalent.

The reader may like to use the ^{19}F n.m.r. spectrum for ClF_3 (two main absorptions at different chemical shifts, in the intensity ratio $2:1$, with splittings of $1:1$ and $1:2:1$ respectively) to determine the shape of the molecule. Note that the chlorine nucleus has no effect on the spectrum and that fluorine–fluorine splittings are governed by the same rules as those deduced in section 5.2 for hydrogen–hydrogen splittings.

Further reading

J. D. Roberts, "Nuclear Magnetic Resonance Spectroscopy", *J. Chemical Education,* 1961, **38,** 581.

M. H. Proffitt and W. C. Gardiner, Jr., "Instructional N.M.R. Instrument", *J. Chemical Education,* 1966, **43,** 152.

More advanced reading

The four books referred to in connection with more advanced reading on i.r., u.v., etc. (pages 69 and 85) all contain useful chapters on more advanced aspects of n.m.r. spectroscopy.

Chapter 6

X-ray, neutron and electron diffraction

The techniques described in this chapter are quite different from the spectroscopic methods so far discussed. Whereas the latter are based on the absorption of certain wavelengths (and therefore certain energies) from radiation comprising a range of wavelengths, the diffraction techniques employ radiation with a single wavelength, i.e. **monochromatic** radiation.

For example, X-ray diffraction occurs when a monochromatic beam of X-radiation interacts with matter and is *scattered* in different directions, with no absorption of energy. Similarly, a beam of neutrons or a beam of electrons with well-defined wavelength (both neutrons and electrons, like the electromagnetic radiation previously referred to, have wave properties) can be scattered to give typical diffraction patterns.

The basis of the application of diffraction techniques in chemical problems is to use ions or molecules as diffraction gratings and then to determine, from the observed diffraction phenomena, the spacings between ions in a crystal or between the atoms which constitute molecules.

6.1 X-RAY DIFFRACTION

The first significant experiments were carried out at the beginning of the present century when it was realised both that X-rays have wave properties and that crystals consist of regular arrays of atoms or ions. It was argued and subsequently demonstrated by von Laue in 1912 that the crystal lattice should behave as a grating and that it should therefore be possible to generate a diffraction pattern from a crystal; for his first experiments he used an ionic crystal, a beam of X-rays (with a range of wavelengths) and a photographic plate (to detect the scattered X-rays). A regular pattern of spots appeared on the plate, giving a clear indication of the success of the experiment. Note that since diffraction phenomena can only be observed if the wavelength of the radiation is smaller than the separation of the atoms or ions in a crystal, X-rays (but not visible light) fulfilled this condition.

This chapter describes how the method has been developed to provide a means for determining the exact positions of ions in an ionic crystal lattice and of atoms within a molecule—that is, for determining accurate values for bond angles and bond lengths, even in extremely complicated molecules like proteins and enzymes.*

(a) X-ray Diffraction Apparatus

X-rays are produced when a beam of accelerated electrons strikes a metal target. An inner electron from an atom in the metal is ejected, an outer electron drops down to fill the vacancy created (see Chapter 2) and the emitted radiation, of precise energy, frequency, and hence wavelength, is in the X-ray region. Since various electronic transitions are possible, the resultant beam at this stage contains X-rays of several different energies (wavelengths). Figure 6.1, for example, shows the intensity of X-radiation, as a function of λ, emitted from a copper target. The K_{α} line, which corresponds to the energy emitted when an electron undergoes a transition from the L shell to the K shell ($2p \rightarrow 1s$ in terms of orbitals) has $\lambda = 0.154$ nm. A sheet of nickel proves to be a good filter for all the wavelengths except the K_{α} line (all the X-rays of wavelength less than the "absorption edge" of 0.149 nm are absorbed: they are of high enough energy to remove completely a K-shell electron from a nickel atom) so the combination of the two metals used like this provides a monochromatic beam of radiation.

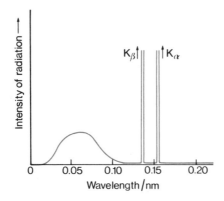

6.1   Intensity of X-radiation at different wavelengths emitted from a copper target

It is perhaps worth mentioning here that other metals can be employed to give X-ray beams with different λ. H. G. Moseley discovered the linear relationship between $\sqrt{\nu}$ (frequency) and Z (atomic number)

---

* "The Start of X-ray Analysis", by Sir Lawrence Bragg (*Chemistry Background Book*, Longmans–Penguin, 1967) and Topic 8, "Chemistry: Students' Book I", *Nuffield Advanced Science*, Penguin, London, 1970, provide an excellent introduction to the subject.

of the K_α line for a variety of elements.* Moseley's work later allowed uncharacterised elements, from which X-rays of a certain frequency were detected, to be given their correct positions in the Periodic Table.

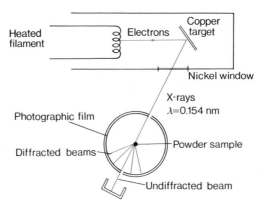

6.2a   Basic features of an X-ray camera for use with powdered samples

The monochromatic beam of X-rays is incident on a solid sample of the material under investigation—either a single crystal, if one is available, or a powder. The powder contains many very small crystals, in a variety of different orientations, whereas for a single crystal only one orientation of the solid can be considered at a time.

Detection of the resultant X-ray beam is usually achieved by surrounding the sample with a photographic film: where X-rays strike the film it becomes darkened, and the film is subsequently developed to yield the diffraction pattern. Figures 6.2a and b show the experimental arrangement for taking a powder "photograph" and Figures 6.3 and 6.4 are the thin strips of film (opened-out) recording the diffraction patterns from two powdered metals (molybdenum and copper, respectively). Figure 6.5 is the diffraction pattern from powdered sodium chloride. The photographs indicate that the X-rays, on striking the powders, become diffracted into a series of well-defined cones. The origin of this phenomenon will be discussed in the next section.

---

* An account of Moseley's work can be found in "Chemistry: Students' Book I", *Nuffield Advanced Science*, Penguin, London, 1970. Data on the wavelengths of the K_α and K_β (*M* to *K* shell) transitions for various elements can be found on p. 53 of "Book of Data: Chemistry, Physical Science, and Physics," *Nuffield Advanced Science*, Penguin, London, 1972; this also includes details of the "absorption edge" for various metals. The reader might like to use the data on ν and *Z* to check that Moseley's relationship holds for the various transitions listed.

6.2b   Arrangement of the film for powder photograph : the resulting "lines" can be seen

If a single crystal is employed, it is usually mounted at the centre of a cylindrical film of somewhat greater depth than that used in the powder method. The crystal is arranged with one of its major axes vertical (this can be achieved after close inspection with a microscope). The diffraction pattern is then recorded, often with simultaneous rotation of the crystal about the axis (or even with rotation or oscillation of both film and crystal). A typical single crystal photograph shows several layers of spots; Figure 6.6 is an X-ray single crystal rotation photograph of sodium chloride. The derivation of information from the photographs will follow the explanation of the origin of the patterns.

(b) The Bragg Equation

A more detailed understanding of X-ray diffraction and of the exact requirements for the appearance of intensity maxima were presented by W. L. Bragg in 1912 (you are encouraged to read his own account—see the footnote on page 121).

Bragg realised that when X-rays impinge on a crystal some are reflected from the atoms in the top layer, whereas others penetrate this layer and are reflected off the next layer, and so on. Exact analysis then

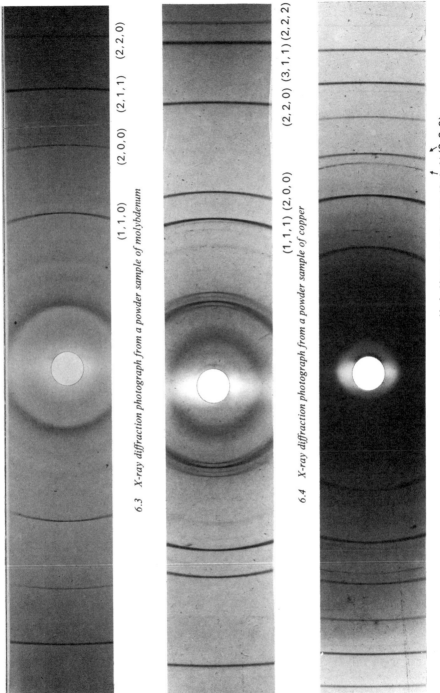

(1, 1, 0)   (2, 0, 0)   (2, 1, 1)   (2, 2, 0)

6.3 X-ray diffraction photograph from a powder sample of molybdenum

(1, 1, 1) (2, 0, 0)   (2, 2, 0) (3, 1, 1) (2, 2, 2)

6.4 X-ray diffraction photograph from a powder sample of copper

(1, 1, 1)(2, 0, 0)(2, 2, 0) (3, 1, 1)(2, 2, 2)

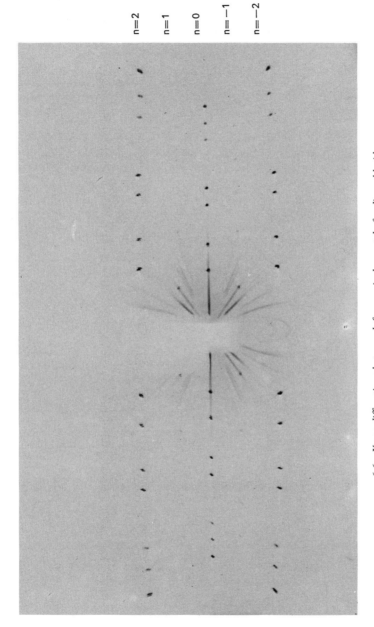

6.6   X-ray diffraction photograph from a single crystal of sodium chloride

showed that the resultant rays are only in phase for certain angles of incidence of the X-rays upon the crystal. This is illustrated in Figure 6.7a, from which it can be seen that there is a path difference between the two rays (one reflected off the top layer, the other reflected off the second layer) when they arrive at the detector. For these two rays to combine (i.e. to reinforce each other) they must be completely in phase. That is, their path difference must be a whole number of wavelengths:

$$n\lambda = 2d\sin\theta \qquad (6.1)$$

where d is the separation between the planes and λ is the wavelength of the X-rays. This is the **Bragg Equation** and it predicts that a reflected beam will be observed (i.e. that there is *constructive* interference) only for certain angles of incidence of the X-ray beam on the crystal; at other angles of incidence the rays from the different layers will be partly or completely out of phase (*destructive* interference).*

For example, if $\lambda = 0.154$ nm and $d = 0.2$ nm, then as θ (the angle of incidence) is steadily increased, reflection will first occur when $n = 1$ and $\sin\theta = 0.385$, i.e. when θ is approximately $23°$. The condition may also be satisfied for $n = 2, 3, 4$, etc., and hence for higher values of θ, but we will normally be concerned with the *first order* ($n = 1$) reflections.

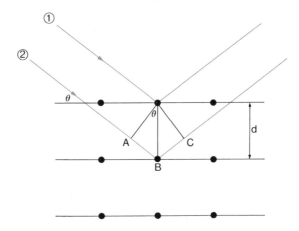

6.7a *Reflection of X-rays from the first and second rows of atoms (ions) in a solid lattice. The path difference between the reflected rays is:*
$$(AB + BC) = 2d\sin\theta,$$
where θ is the angle of incidence and d is the distance between the planes

---

*A simple Bragg phase difference simulator is described in pp. 267–269 of "Chemistry: Teachers' Guide II", *Nuffield Advanced Science*, Penguin, London, 1970.

Figure 6.7b illustrates one particular small crystal (crystallite) in a *powder*, oriented with its surface plane of atoms at an angle θ to the beam. If this value of θ fulfils the Bragg equation for the particular values of λ and d in the experiment (since there are many crystallites in the powder, this will be true for some of them), then the beam is reflected to the film. There will be other crystallites each making the angle shown, which means that a cone of diffracted X-radiation will be produced, darkening the film at the points indicated (see also Figures 6.2–6.5). A *series* of cones is produced because there are various possible spacings in the crystal (with different values of d) which satisfy equation 6.1 for different values of θ.

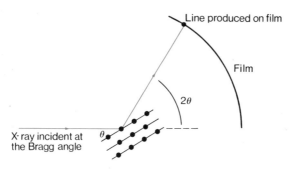

6.7b   *Production of a "line" on the photographic film from diffraction of X-rays incident at the Bragg angle*

When a *single crystal* is used, with one axis vertical, then the regular inter-atomic spacing along this axis behaves as a simple diffraction grating (Figure 6.8). Thus the layers of lines observed (see Figure 6.6)

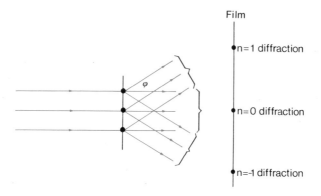

6.8   *Diagrammatic representation of diffraction by a single crystal mounted with one axis vertical : the diffractions with n = 1, 0, − 1 etc., correspond to the layers of lines clearly visible in Figure 6.6*

are simply the reflections which satisfy the following equation for λ, d (the vertical spacing, Figure 6.8) and $n = 0, 1, 2$, etc:

$$n\lambda = d \sin \varphi \qquad (6.2)$$

The layer lines themselves are clearly not continuous, because for some points on the layer lines there will be destructive interference between the reflections which satisfy equation 6.2 for the vertical spacing and reflections from other planes in the crystal. To understand this and to appreciate the relationship between the possible spacings in a crystal (and hence all the possible values of d) and the structure of the lattice, a little familiarity with crystal types and nomenclature is needed.

6.2 CRYSTALLOGRAPHY

(a) Unit cells and crystal systems

A crystal consists of a repeating **unit cell** of atoms or ions, in three dimensions. The unit cell is characterised by the length of each side and the angles between the three sides. Figure 6.9 shows three of the possible different types of unit cell—these correspond to three of the seven **crystal systems**. The symmetry of each of these allows three-dimensional structures (crystals) to be built up from the tiny building blocks (unit cells). The planes referred to earlier are sheets in the crystal containing a high density of lattice points (atoms or ions). These planes can be observed as the external faces of a crystal.

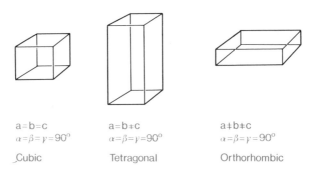

a=b=c	a=b≠c	a≠b≠c
$\alpha=\beta=\gamma=90^\circ$	$\alpha=\beta=\gamma=90^\circ$	$\alpha=\beta=\gamma=90^\circ$
Cubic	Tetragonal	Orthorhombic

6.9 *Three of the possible crystal systems*

Within the unit cell there are various allowed arrangements of atoms or ions which still preserve the overall symmetry of the cell. Figure 6.10 shows the three possibilities for a **cubic** unit cell (i.e. with equal sides, all angles 90°); these are the **primitive** (or **simple**) cubic, **body-**

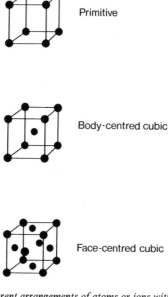

Primitive

Body-centred cubic

Face-centred cubic

6.10   Possible different arrangements of atoms or ions within a cubic unit cell

centred cubic, and **face-centred** cubic systems. There are fourteen different lattices (called the Bravais lattices) which arise from the seven crystal systems.

At this stage the reader will probably find it helpful to inspect "ball-and-stick" models of unit cells and fragments of lattices in order to appreciate the various possible planes of atoms or ions in the different types of crystal.

Further it can then also be recognised that a face-centred cubic lattice can equally well be described as a cubic-close packed arrangement; this common structure arises if equivalent spheres (e.g. styrofoam balls) are packed in one layer, the next layer is added, and then the third layer is added at the positions *not* over the first layer but in the alternative positions; the next layer goes over the first, to give an A B C A B C stacking pattern.

(b) Lattices in solids

The lattices we are describing can be of various chemical types:

(i) *Atoms (metals or alloys)*. For example, metallic caesium exists in a body-centred cubic lattice, as does α-iron, whereas copper has a face-centred cubic pattern. For a metal there is only one type of atom in the unit cell.

(ii) *Ionic crystals*. These contain two or more different types of ion (e.g. caesium chloride, sodium chloride), and they can be characterised in the same way as the simple lattices. For example, caesium chloride consists of a simple cubic array of Cs^+ ions with Cl^- ions at the centre of each cube. The reader should be able to demonstrate that this places 8 Cs^+ ions round each Cl^- ion and *vice-versa*. The arrangement is referred to as two inter-penetrating primitive cubic lattices. For sodium chloride the crystal structure is of two inter-penetrating face-centred cubic lattices: the complete unit cell is shown in Figure 6.11.

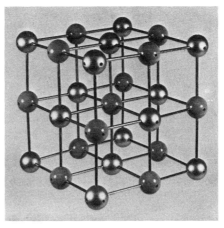

6.11   Model of the unit cell of sodium chloride, NaCl. One type of ion (Na^+ or Cl^-) can be seen at the corners of the cube and at the centre of each face: the other ion occupies vacancies created in this lattice and itself has a Face-Centred Cubic structure. Model by Crystal Structures Ltd.

It must be remembered that the drawing of small circles and the use of small balls and "bonds" for models is essentially for convenience in visualising the planes and unit cells in the lattice. Close-packing of larger space-filling spheres produces a more realistic model in which the electron clouds of neighbouring ions are seen to be in contact.

(iii) *Covalent molecules.* Covalent compounds also form crystals, with *molecules* at the lattice points in the unit cells, exactly as those for atoms and ions.

(c) Planes in the crystal

It is desirable to adopt a shorthand procedure for describing a particular plane in the crystal, and the system of **Miller Indices** is universally employed. Their definition can be illustrated for the three-dimensional lattice illustrated in cross-section in Figure 6.12; the x and y axes are indicated, with the z-axis coming out of the paper. Each dot then represents a vertical column of atoms. The three lines indicated are three planes we wish to describe (these are just three

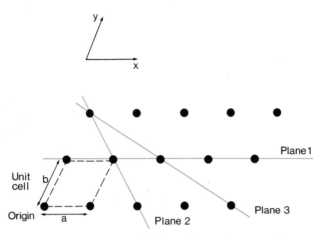

6.12   Representation of different planes within a crystal

of the many possible planes). Miller Indices provide a short-hand means of referring to any one of these and, equally, to all of those parallel to any one we choose.

The procedure is as follows:

(i) choose an origin;
(ii) read off the intercepts along the x, y and z axes in terms of the unit spacings in the crystal (a along the x-axis, b along the y-axis, c along the z-axis). The intercept is infinity if the plane does not cut the axis concerned;
(iii) take reciprocals, dropping any reference to a, b and c;
(iv) if the reciprocals for a given plane include a fraction or fractions, express the reciprocals as their simple integral ratio, leaving Miller Indices, which are then referred to as h, k, l values.

Step (i)	(ii)	(iii)			(iv)
Plane	*Intercepts*	*Reciprocals*			Miller Indices h, k, l
(1)	∞a, $1b$, ∞c	$1/\infty$	$1/1$	$1/\infty$	(0, 1, 0)
(2)	$2a$, $2b$, ∞c	$1/2$	$1/2$	$1/\infty$	(1, 1, 0)
(3)	$4a$, $2b$, ∞c	$1/4$	$1/2$	$1/\infty$	(1, 2, 0)

Thus plane *(1)* and *all those parallel to it* are referred to as (0, 1, 0) planes; similarly plane *(2)* and those parallel to it are called (1, 1, 0) planes. The reader should check that it does not matter where the origin is taken.

To recapitulate, a, b and c are properties of the unit cell, as are the angles between the axes; h, k and l are integers used to describe any particular plane in the crystal and, as will be seen, they are used to calculate the distance between the planes (i.e. d in the Bragg equation). Incidentally, a 3-dimensional lattice can cleave (or grow) along any of these planes, which accounts for the external forms of crystalline compounds.

At this stage you are encouraged to classify some planes to become familiar with the nomenclature. Given the following planes and axis system, (i) what are the Miller Indices for each plane, and (ii) can you draw the $(0,0,2)$ plane?

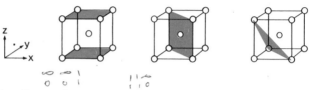

To visualise a plane running through the crystal it is perhaps best to consider several unit cells together or, even better, to refer to a model.

The answers for the planes shown above are, from left to right, $(0,0,1)$, $(1,1,0)$, and $(1,1,1)$ respectively. It should be noted that the $(1,0,0)$ and $(0,1,0)$ planes will be simply related to the $(0,0,1)$ plane— i.e. they are all the appropriate "ends" of the cube. Similarly, $(1,0,1)$ and $(0,1,1)$ will be diagonal planes (like $1,1,0$). The $(0,0,2)$ planes are as follows:

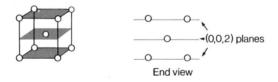

End view

To check this, note that the intercepts are ∞a, ∞b, $\frac{1}{2}c$, with reciprocals $(0,0,2)$.

It is often necessary to describe the perpendicular distance d_{hkl} between parallel planes (h,k,l) of a unit cell with lengths a, b and c. This is the interplanar spacing referred to in the Bragg equation. With some relatively simple geometry it can be shown that for an **orthogonal** crystal system (a unit cell with all angles 90°)

$$\frac{1}{d^2_{hkl}} = \frac{h^2}{a^2} + \frac{k^2}{b^2} + \frac{l^2}{c^2}$$

and, in particular, for a **cubic** system (all angles 90°, and $a = b = c$)

$$\frac{1}{d^2_{hkl}} = \frac{h^2 + k^2 + l^2}{a^2}$$

$$\therefore d_{hkl} = \frac{a}{\sqrt{h^2 + k^2 + l^2}} \quad (6.3)$$

For example, the distance between the $(0,0,2)$ planes in a cubic lattice is simply:

$$d_{002} = \frac{a}{\sqrt{0^2 + 0^2 + 2^2}} = \frac{a}{2}$$

Spacings for other planes in the lattice can be calculated similarly.

6.3 DETERMINATION OF STRUCTURE

(a) Unit cell type and dimensions

X-ray data from a given ionic crystal—in terms of the particular values of θ at which reflections are observed—can be used to determine the type of unit cell in the compound. The Bragg equation [for the allowed values of θ (angle of incidence) for given λ and d] is combined with the expression for the possible values of d in terms of the cell dimensions (a,b,c) of the particular structure considered. For a cubic lattice, for which the spacing y_{hkl} between any set of planes (h,k,l) is given by equation 6.3, the resulting expression which gives the angles of reflection from different planes is derived as follows:

$$n\lambda = 2d \sin \theta,$$

$$\therefore \sin^2\theta = \frac{n^2\lambda^2}{4d^2}$$

$$\therefore \sin^2\theta = \frac{n^2\lambda^2}{4a^2}(h^2 + k^2 + l^2) \quad (6.4)$$

where a is the length of the cell side.

There are many lines in the diffraction pattern (i.e. at different values of θ) because of the various possible values of d. Since for any set of planes h, k, and l must all be integers, then so must $(h^2+k^2+l^2)$; the possible values of h, k, l and $(h^2+k^2+l^2)$ are as follows:

plane h,k,l	1,0,0	1,1,0	1,1,1	2,0,0	2,1,0	2,1,1	2,2,0	2,2,1 3,0,0	3,1,0
$(h^2 + k^2 + l^2)$	1	2	3	4	5	6	8	9	10

Thus for a cubic lattice the increasing values of $\sin^2\theta$ should be related to each other as are the increasing simple integers [the smallest value of $\sin^2\theta$ being the reflection from the $(1,0,0)$ planes, the next from the $(1,1,0)$ planes, etc.] but with no line corresponding to the integer 7 because no combination of the squares of integers gives this number. Similarly it can be shown that there should be no lines for 15, 23, 28: i.e., there is no plane with integers such that $h^2 + k^2 + l^2$ can be 7, 15, 23, 28. The pattern obtained from a simple cubic lattice confirms this analysis, and any solid which gives such a pattern (i.e. with 7th, 15th, etc., lines missing) is known to have a cubic lattice.

Further, if λ is known, values of $\sin^2\theta$ for the $n = 1$ reflections lead to a measurement of a, the length of the side of the unit cell. The process just described is called **indexing** a powder photograph (that is, deriving the shape and size of the unit cell).

The practice is somewhat more complicated if the lattice is of lower symmetry but, once information about a variety of lattices has been accumulated, a diffraction pattern from an unknown type can usually be recognised by comparison with those from known examples.

In the single crystal method, measurements with each axis vertical in turn lead to determination of the unit cell length along each of the axes (from the separation of the layer lines, see page 127). Again, comparison with spectra from compounds whose structures are known can be helpful.

(b) The Bravais lattice

The type of Bravais lattice (e.g. body- or face-centred) can also be readily obtained from the diffraction pattern.

First, consider a body-centred cubic structure and try to visualise what happens to the expected reflections from the $(1,0,0)$ [and the $(0,0,1)$, $(0,1,0)$ planes]. For the $(1,0,0)$ planes, viewed end-on, it can be seen that the reflection which satisfies the Bragg equation (i.e. with distance d_{100}) now has an extra beam superimposed. This is the beam reflected from the atoms at the centres of the unit cells:

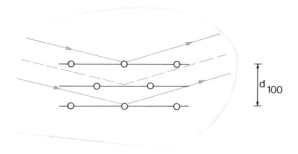

This extra beam is completely *out of phase* with the beams reflected from the $(1,0,0)$ planes (the path difference for the extra beam compared to the others is $\lambda/2$) and therefore there will be destructive interference; the reflection is now *absent*. This is referred to as a **systematic absence.**

The $(2,0,0)$, $(0,2,0)$ and $(0,0,2)$ reflections (with separation d_{200}) will be present, the reflections being in phase (remember that d is now half that for the $1,0,0$ planes); a diffracted beam at the corresponding θ is observed. As Figure 6.13 demonstrates the $(1,1,0)$ reflection should be *present* [there being no atoms in between the $(1,1,0)$ planes] but the $(1,1,1)$ plane reflections are absent (the planes have atoms in between).

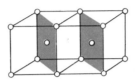

(1,1,0) planes

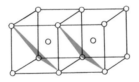

(1,1,1) planes

6.13 *A body-centred cubic structure showing planes with interstitial ions (1,1,1) and planes without interstitial ions (1,1,0)*

For a face-centred cubic structure, the $(1,0,0)$ reflections are absent, as too are the $(1,1,0)$ reflections, but not, in this case, the $(1,1,1)$ type (you are recommended to check that this is the case by reference to models).

The following diagram summarises the permitted reflections:

Cell type	(h,k,l)	1,0,0	1,1,0	1,1,1	2,0,0	2,1,0	2,1,1	2,2,0
Primitive cubic		✓	✓	✓	✓	✓	✓	✓
Body-centred cubic		✓		✓		✓		✓
Face-centred cubic				✓	✓			✓

A simple (primitive) cubic cell gives no systematic absences, whereas the body-centred cubic cell shows only those reflections for which

($h + k + l$) is even; the face-centred cubic structure gives reflections only from planes where h, k and l are either all odd or all even. Clearly this information enables the operator to use a series of values of $\sin^2\theta$ (from a powder photograph) to decide to which type of lattice the numbers belong. For example, if the values of θ are such that the ratios of $\sin^2\theta$ for the first three reflections fit most closely the integers $3 : 4 : 8$, then the lattice is of *face-centred cubic* type [3 has h, k, l, of $(1,1,1,)$ type, 4 has h, k, l of $(2,0,0)$ type, and 8 has h, k, l $(2,2,0)$ type]. The ratios are quite different for primitive or body-centred cubic structures.

This procedure can be exemplified by reference to Figures 6.3 and 6.4; molybdenum has a body-centred cubic structure, whereas copper is face-centred cubic.

All the powder photographs described here were recorded with an X-ray camera employing copper K_α radiation (λ 0.154 nm) and a circular photograph in which 180° corresponds to 180 mm of film: the photographs have been reproduced exactly to scale so that the angle between the two points of each cone (where it cuts the film) can be measured from the figures (1 mm = 1°). This angle will equal 4θ, where θ is the angle of incidence of the beam on the particular plane of the crystal causing diffraction—similarly the angle between the cone and the centre of the film is 2θ, (see Figure 6.7b).

Table 6.1 shows the measured values of θ for copper (Figure 6.4); they have been converted into $\sin^2\theta$ values which are then found to be approximately in the integral ratios 3, 4, 8, 11, corresponding to the solutions to equation (6.4) with h, k, l values, respectively, $(1, 1, 1)$, $(2,0,0)$, $(2,2,0)$ and $(3,1,1)$ as indexed. This is then a *face-centred* lattice. The pattern is very different from the diffraction spectrum of the body-centred molybdenum lattice (indexed on the photograph).

TABLE 6.1

X-ray Powder Diffraction Photograph for Copper

θ/degrees (taken from* the photograph)	$\sin\theta$	$\sin^2\theta$	Integral ratios $(h^2+k^2+l^2)$	h,k,l
22.0	0.3746	0.1403	3	1,1,1
25.5	0.4305	0.1853	4	2,0,0
37.3	0.6060	0.3672	8	2,2,0
45.0	0.7071	0.5000	11	3,1,1
47.8	0.7408	0.5488	12	2,2,2

*See Figure 6.4. Angle between line and centre of film corresponds to 2θ: 1 mm = 1°.

From any one of the values of θ (and the associated h, k, l) from Table 6.1, together with the known value of λ, a can be calculated using equation 6.4 (remember that $n = 1$). In this example the answer for a (the length of the side of the unit cell in copper) is 0.356 nm.

(c) Ionic Crystals

The powder photograph of crystalline sodium chloride provides a good example of the effect upon the diffraction pattern when more than one type of ion is present in the lattice.

Table 6.2 shows the values of θ and $\sin^2\theta$ for this crystal (recorded with $\lambda = 0.154$ nm). The values of $\sin^2\theta$ are indexed in terms of h, k and l, for the appropriate planes, and the results provide confirmation of a face-centred cubic structure (see page 135). Further, from equation 6.4 it is possible to use a given set of h, k, l values and the appropriate values of $\sin^2\theta$ to derive a, the length of the side of the unit cell.

TABLE 6.2

X-ray Powder Diffraction Photograph for Sodium Chloride

θ/degrees*	$\sin\theta$	$\sin^2\theta$	Integral ratios ($h^2 + k^2 + l^2$)	h,k,l
13.9	0.2402	0.0586	3	1,1,1
16.0	0.2756	0.0759	4	2,0,0
22.8	0.3875	0.1502	8	2,2,0
27.0	0.4540	0.2061	11	3,1,1

* See Figure 6.5. The angle between a line and the centre of film corresponds to 2θ; 1 mm $= 1°$. The weak lines associated with the reflections from 1,1,1 and 3,1,1 planes are almost invisible here: see text.

However, it can be recognised from the diffraction pattern that the $(1,1,1)$ reflections are weaker than those for the $(2,0,0)$ planes. This observation can be understood in terms of the intensity of the scattered beam of X-rays and its dependence on the nature of the atom doing the scattering. The X-rays are scattered by the electrons around the nucleus, which will lead to a different intensity (amount) of scattering from Na^+ and Cl^-. Now for the $(2,0,0)$ type reflections, all the planes causing reflection contain Cl^- and Na^+ ions so that all these ions will contribute to reinforcement [three $(2,0,0)$ horizontal layers are clearly shown in Figure 6.11]. However, for the $(1,1,1)$ plane of sodium ions (Figure 6.14 shows the unit cell tipped on to one corner to illustrate this), there is a layer of chloride ions in between; the X-rays scattered

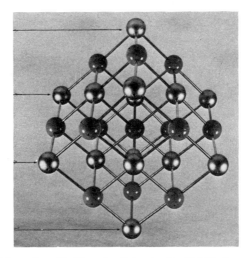

6.14   View of the sodium chloride unit cell showing the (1,1,1) planes of one type of ion (indicated) and interstitial planes comprising ions of the other type. Model by Crystal Structures Ltd.

from each layer are exactly out of phase, but do not cancel exactly because their *intensity* is not the same. Thus a weak reflection is observed.

Confirmation of this conclusion results when a crystal where the two ions have the same number of electrons is studied. This is true, for example, for KCl; these two ions, with the same number of electrons, cannot be distinguished with X-rays. Although the KCl structure is the same as that for NaCl, the (1, 1, 1) reflection is now actually missing. The X-ray pattern resembles that for a cubic lattice with one half of the spacing of the actual unit cell.

Thus the important information in an X-ray diffraction pattern is contained in the *positions* (values of θ) and *intensities* of the diffracted beams. The former information usually allows recognition of the unit cell type, the Bravais lattice and the dimensions of the unit cell. The intensities of the reflections depend upon the *nature* of the ions present and these hold the key to the application of X-ray diffraction for structure determination in complex organic molecules (see p. 140).

6.4   WORKED EXAMPLES

(a) Problems
(i) When X-rays of wavelength 5.81×10^{-11} m impinge on a powdered sample of NaCl, the $(2,0,0)$ reflections occur at $\theta = 5.9°$.

Calculate the length of the side of the unit cell (this is a face-centred cubic structure).

(ii) Using the result from (i) and, given the value of the Avogadro constant $(6 \times 10^{23} \text{ mol}^{-1})$ calculate the density of NaCl (the observed value is $2.163 \times 10^3 \text{ kg m}^{-3}$).

(iii) Predict the next value of $(h^2 + k^2 + l^2)$, and the associated values of (h, k, l), θ, and $\sin^2\theta$, for the crystal (NaCl) for which data is given in Table 6.2. Would you expect this to be a strong or a weak reflection?

(b) Discussion
(i) Clearly, to calculate the cell side (a) from θ we must employ the Bragg equation,

$$n\lambda = 2d \sin\theta$$

taking n $= 1$, $\lambda = 5.81 \times 10^{-11}$ m, $d = a/2$ [since the $(2,0,0)$ layers are half the cell side apart (see page 132)], and $\theta = 5.9°$,

$$a = (5.81 \times 10^{-11})/\sin 5.9° = 0.56 \text{ nm}.$$

(ii) The volume of the unit cell is $(0.56 \times 10^{-9})^3$ m³ and each unit cell contains, effectively, four Na^+ and four Cl^- ions. This can be verified by inspection of the structure (Figure 6.11); for book-keeping purposes, count $\frac{1}{2}$ for each ion shared between two unit cells (i.e. for ions in the middle of the sides) and $\frac{1}{8}$ for each ion at the corners (shared between eight cells).

Mass of the unit cell $= 4$ (mass of Na^+ ion + mass of Cl^- ion)

$$= \frac{4(\text{relative atomic mass of } Na^+ + \text{relative atomic mass of } Cl^-)}{\text{The Avogadro constant}}$$

$$= \frac{4 \times (23 + 35.5)}{6 \times 10^{23}}$$

$$= 39 \times 10^{-23} \text{ g}$$

$$= 39 \times 10^{-26} \text{ kg}$$

Density of NaCl $= \dfrac{39 \times 10^{-26}}{(0.56 \times 10^{-9})^3}$

$$= 2.2 \times 10^3 \text{ kg m}^{-3}$$

In practice, the density would be known, so that the calculation would be done in reverse—i.e. to confirm the number of atoms or ions in the unit cell.

(iii) The next allowed reflection is $(2,2,2)$ with $(h^2 + k^2 + l^2) = 12$. Then:

$$\sin^2\theta = 0.2061 \times 12/11 = 0.2250,$$
$$\theta = 28.30°$$

This should be a *strong* reflection, since rays from layers of Na^+ ions and from layers of Cl^- ions will be in phase. Inspection of Figure 6.5 confirms the expected pair of lines corresponding to this value of θ.

6.6 FURTHER APPLICATIONS

(a) Ionic Radii

It has been shown how X-ray diffraction enables different types of lattice to be distinguished. It is found that even within a group of closely related compounds different structures can exist (e.g. the sodium chloride and caesium chloride types for alkali halides). This information is vital to the understanding of the effect of relative sizes and charges of the ions on the packing of ions to form a crystal.

Further, the measurement of the unit cell length can lead to values of the ionic radii of the constituent ions. For example, for NaCl the length of the cell side is equal to the sum of the diameters of the sodium and chloride ions. In LiCl, which has the same type of structure, the Li^+ ions are so much smaller than the Cl^- ions that the latter actually "touch" along the diagonal of a face (see, e.g., Figure 6.11). From the X-ray powder photograph from this compound, a measurement of the length of the cell side can be obtained and used to calculate the length of the diagonal of the face and hence the ionic radius of Cl^-. Then, from the length of the cell side for NaCl, a measure of the ionic radius of Na^+ can be obtained.

(b) Molecular Structure

In a crystal where the lattice points are not occupied by atoms or ions but by covalent molecules (e.g. in a crystalline organic compound) the diffraction from each separate atom must be considered. The molecules themselves will be symmetrically placed with respect to each other, but atoms in the molecules will now not only be found at the corners and body- or face-centres of the unit cell. However, there will still be characteristic "repeat distances" in the structure (i.e. periodic variation in electron density) which cause diffraction as discussed before for simple ions, and the essential theory of the method is the same. The X-ray examination is carried out for a single crystal, if one is available, and a very complicated pattern of reflections, with different intensities, is obtained. The problem then is to work back from this information to a plot of the electron density (which causes diffraction) in the unit cell.* An electron density contour map, in which the "peaks" correspond to atoms in the unit cell, can be drawn.

---

* A pictorial description is given by J. Waser, *J. Chemical Education*, 1968, **45**, 446.

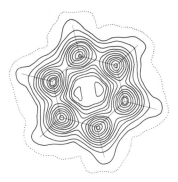

6.15   Electron density map for benzene, derived from a single crystal X-ray diffraction analysis (reproduced from the paper by E. G. Cox, D. W. J. Cruickshank and J. A. S. Smith, Proceedings of the Royal Society, 1958, 247A, p. 1)

Figure 6.15 shows an electron density map for benzene, and the positions of the carbon atoms (a symmetrical hexagon, with equal C—C bond lengths) can be seen: peaks from hydrogen atoms are not clearly observed because there is relatively little X-ray scattering from these atoms (scattering depends on the number of electrons round a given atom: see page 137).

A modern computer makes this type of analysis much easier than was previously possible, and, with this technique, complicated molecules have been investigated and their atomic co-ordinates found. This method then provides accurate measurements of bond lengths and angles for compounds in solids.*

(c) **Natural and man-made polymers**

X-ray diffraction provides an excellent method for investigating the structures of biologically-important molecules which contain repeating chemical groups (e.g. proteins and nucleic acids).

One notable example is the analysis of the X-ray spectrum from the nucleic acid DNA, which is interpretable in terms of this molecule having an interwoven double-helix of repeating units. X-ray reflections indicate characteristic "repeat distances" (cf. planes in a crystal) of 0.34, 3.4 and 2.0 nm; these are, respectively, the distance between successive units in the chain, the repeat distance (pitch of the helix) and the width of the spiral. This information enabled the essential features of the structure to be elucidated. It was eventually possible

---

*Further examples of electron density maps appear on pp. 237 and 239 of "Chemistry: Students' Book I", *Nuffield Advanced Science*, Penguin, London, 1970 and on pp. 232 and 250 of "Chemistry: Students' Book II", *Nuffield Advanced Science*, Penguin, London, 1970.

to obtain an electron-density map indicating the detailed structure; examples of an X-ray photograph, electron density diagram, and molecular model of DNA are given on pages 215–217 of *Chemistry; Students' Book I* (referred to on page 141).

Applications of X-ray diffraction to the study of man-made fibres allows a distinction to be made between polymers with a regularly repeating structural pattern and those which lack this degree of order. Further references are given at the end of the chapter.

6.6 PROBLEMS

6.1 Figure 6.3 shows the X-ray powder photograph from molybdenum (recorded with X-rays of wavelength 0.154 nm); the film has $1° = 1$ mm and is reproduced to scale. From the positions of the $(2,0,0)$ reflections calculate:
   (a) the Bragg angle for this reflection
   (b) the length of the side of the unit cell for molybdenum.

6.2 Figure 6.6 is the single crystal photograph (to scale) obtained using X-rays of wavelength 0.154 nm from a crystal of sodium chloride with one axis vertical. Figure 6.8 and equation 6.2 describe how the layer lines arise. The angle φ for any particular layer line can be obtained as follows:

$$\tan \varphi = \frac{y}{R}$$

where R is the radius of the film (29 mm in this example) and y is the vertical distance between the layer line and the $n = 0$ layer.
   (a) Calculate φ for the $n = 1$ layer and hence calculate the value of a, the length of side of the unit cell.
   (b) Why are the spots in the $n = 1$ and $n = -1$ layers weaker than those in the $n = 2$ and $n = -2$ layers?

6.3 (a) Lithium chloride has a sodium chloride-type lattice, with the length of the side of the unit cell 0.513 nm. On the assumption that the chloride ions actually "touch" along the diagonal of a face of the unit cell, derive a value for the ionic radius of Cl^-. Use this value together with the length of the cell side for NaCl (0.56 nm) to derive the ionic radius of Na^+.

   (b) The powder photograph of LiCl has reflections at increasing angles (θ) which have values of $\sin^2\theta$ in the ratio 3, 4, 8, 11, 12, etc. (like NaCl). Unlike NaCl, for which the reflections associated with some of these (3 and 11) are much weaker than the others, all the reflections for LiCl have approximately the same intensity. Can you account for this observation?

6.7   NEUTRON DIFFRACTION

One of the disadvantages of X-ray diffraction is that atoms of low atomic number are difficult to locate in the present of heavier elements (in particular, hydrogen atoms are difficult to detect even in the presence of carbon or oxygen) because the intensity of scatter (or "reflection") from an atom depends on the number of electrons around that atom.

If a study of the precise positions of hydrogen atoms is to be made (e.g. a study of a metal hydride or a hydrated salt) then the related method of **neutron diffraction** can be employed. In this technique a beam of neutrons (from an atomic pile) is diffracted in exactly the same manner as is a beam of X-rays, and the intensity of the beam scattered in different directions is monitored. There is much less variation with atomic number in the scattering of neutrons than there is for X-rays, so that the technique becomes a more sensitive probe for hydrogen atoms than is X-ray diffraction. Obviously neutron diffraction is both difficult and costly, and it is only usually attempted when a full X-ray analysis (to locate all the heavier atoms) has been performed.

6.8   ELECTRON DIFFRACTION

At approximately the same time that X-ray diffraction was first demonstrated, other investigators showed that electrons possess wave-character and that they too can exhibit interference patterns.

For example, it was shown that when an electron beam is fired through a very thin gold foil, a diffraction pattern consisting of concentric dark rings appears on a photographic plate behind the metal. This type of experiment does not find wide application, because electrons tend to be absorbed by, rather than to be scattered by, denser forms of matter (liquids, solids). For this reason although some work is carried out with thin films, very small crystals, and surfaces (see page 146) most of the electron diffraction applications in the field of molecular structure have been concerned with gaseous samples.

(a)   Theory and Instrumentation

The sample under investigation is usually examined at low pressure (*ca.* 10^{-4}N m^{-2}) and the incoming electron beam is scattered by the electric field of the atoms, provided by the nuclei and electrons.

A single atom scatters the electron beam in all directions; however, for a molecule the rays scattered from the constituent atoms (see, e.g. Figure 6.16) will be in phase only in certain directions. These will

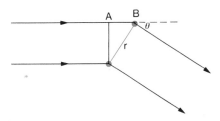

6.16   Diffraction of an electron beam by a diatomic molecule

depend (cf. the Bragg equation) on λ (for the electrons) and the path difference, d, which itself depends on the bond length and the angle made by the bond to the direction of the beam. For a large collection of such molecules there will be many molecules at the correct angle θ for in-phase reflection, so that a beam which consists of a cone of diffracted electrons is produced (as with X-ray diffraction from powders). There will be more than one angle at which this occurs, and the result is recorded on a photographic film as a series of dark and light rings (see Figure 6.17).

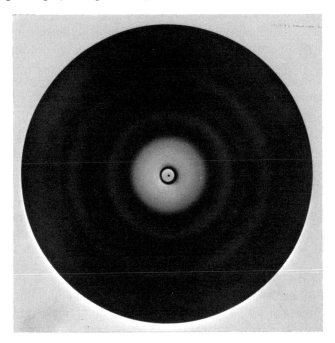

6.17   Electron diffraction pattern from tetrafluoroethene

The beam of electrons must have a wavelength somewhat smaller than typical bond lengths if diffraction is to be detected (as with X-rays).

This is arranged as follows. The wavelength of an electron depends on the mass and the velocity of the electron (the **de Broglie** equation, 6.5)

$$\lambda = \frac{h}{mv} \qquad (6.5)$$

The velocity of the electrons can be altered using an accelerating potential V (see also page 10), so that

$$\tfrac{1}{2} mv^2 = eV \qquad (6.6)$$

Combining these two equations it follows that

$$\lambda = \frac{h}{m} \sqrt{\frac{m}{2eV}} \qquad (6.7)$$

With $V = 40$ kV, λ is *ca.* 0.006 nm, and this proves suitable for electron diffraction studies.

(b) Structure Determination

Gas-phase electron diffraction is normally carried out only for small molecules. The necessity for this simplification arises because the separations (see Figure 6.16) which each give rise to diffracted "cones" are those between bonded *and* between non-bonded pairs of atoms in a molecule.

Thus, for CCl_4 for example, there will be scattering at certain angles from the C—Cl bonds but also from the Cl·····Cl pairs in the same molecule (i.e. these pairs also behave as a grating). From the number and spacing of the rings in this fairly simple example it is possible to deduce that the molecule is *tetrahedral* (rather than square planar) and that the C—Cl bond length is 0.177 nm.

However, if the molecule under investigation is a little more complicated then there are more separations to confuse the analysis. For CF_3Cl, for instance, the distances responsible for diffraction are C—Cl, C—F, F·····Cl and F ·····F . A detailed solution must also allow for the different types of atom (different scattering intensities), and in practice it can be carried out to give bond lengths to about 0.001 nm and to confirm the molecular shape. In this respect the results are similar to those obtained using infra-red spectroscopy.

The method can be extended to more complicated molecules (though electron diffraction is rarely, if ever, used as a service instrument for routine structure determination). For example, the electron diffraction pattern for benzene provides evidence for a molecule with three separate interatomic C·····C distances and four interatomic C·····H distances, consistent with a regular planar hexagonal arrangement of the six carbon atoms in the molecule (all C—C bonds the same length, 0.139 nm). In contrast, the cyclooctatetraene molecule (C_8H_8), can be shown to be tub-shaped, with alternating long and short bonds:

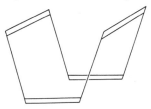

(c) Low Energy Electron Diffraction
This method can be employed to probe the arrangement of atoms in a solid surface; electrons with a wavelength in the range 0.5–0.05 nm (and hence with energy lower than that associated with gas-phase electron diffraction) are directed on to the sample (e.g. a metal). Electrons are reflected from the top layer of atoms, and the diffracted in-phase beam in certain directions (cf. the Bragg equation) is detected on a fluorescent screen. The resulting pattern leads to details of the two-dimensional structure of the top layer.

The method can be applied to the study of the adsorption of gases on solids, and in this way it has provided a useful technique for recent research on the mechanisms by which some solids act as catalysts.

6.9   SUMMARY
The diffraction techniques discussed in this chapter all involve complicated equipment, difficulties in handling and, in most cases, computer processing of data (except for the analysis of the X-ray diffraction spectra of simple atomic or ionic lattices). They are clearly capable of solving detailed problems of molecular structure, but should be contrasted with the more routine spectroscopic techniques discussed earlier.

The two types of technique often have fairly different applications. The spectroscopic techniques have the advantage of giving a rapid indication of the various groups in a molecule and the way in which they are joined together. This type of analysis is very widely used, especially in conjunction with parallel investigations of chemical reactions which diagnose the presence of functional groups and their relative positions in the molecules concerned. However, for a solid sample an X-ray diffraction study (and possibly even a neutron diffraction study) may be later pursued in order to confirm a proposed structure and to measure more detailed molecular parameters, e.g. bond lengths and angles; for a compound which is readily vaporised an electron diffraction experiment might similarly be undertaken.

Further reading

Sir Lawrence Bragg, "X-ray Crystallography", *Scientific American*, 1968, **219**, July, p. 58.

P. J. Wheatley, "The Determination of Molecular Structure", 2nd Edn., Oxford, 1968.

D. O. Hughes and E. D. Morgan, "A student exercise in X-ray crystallography", *Education in Chemistry*, 1970, **7**, 59.

X-ray Spectra of Ionic Crystals. X-ray powder spectra from lattices of different type are represented in the article "X-ray Crystallography Experiment: Powder Patterns for Alkali Halides" by F. P. Boer and T. H. Jordan, *J. Chemical Education*, 1965, **42,** 76. The spectra clearly illustrate the dependence of the positions and intensities of the reflections on the lattice type and the nature (size and electron density) of the constituent ions. See also the article by N. Booth, "X-rays and the alkali-metal chlorides", *Education in Chemistry*, 1964, **1**, 66.

Use of X-rays for determining the structures of naturally occurring compounds: "Chemistry: Students' Book II", *Nuffield Advanced Science*, Penguin, London, 1970, pp. 231–254 and 266–278 (proteins, polypeptides, enzymes, polymers and description of an optical analogue of diffraction from helical structures).

J. C. Kendrew, "The Three-Dimensional Structure of a Protein Molecule", *Scientific American*, 1961, **205,** December, p. 96.

D. C. Phillips, "The Three-Dimensional Structure of an Enzyme Molecule", *Scientific American*, 1966, **215,** November, p. 78.

Additional Problems

7.1 A compound, X, has the composition 62.0% C, 27.6% O, and 10.4% H. Its n.m.r. and i.r. spectra are shown in Figures 7.1a and 7.1b, respectively. The u.v. spectrum of X (recorded for a solution in hexane) has $\lambda_{max} = 290$ nm, $\varepsilon = 1.8$ m^2 mol^{-1}, and the ten most abundant peaks in the mass spectrum are as follows:

m/e	29	58	28	27	57	18	41	39	15	55
relative abundance/%	100	83	82	57	26	8	7	6	5	4

Identify compound X.

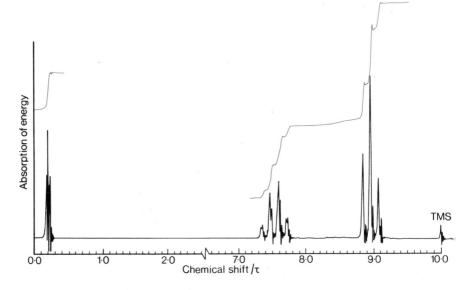

7.1a   N.m.r. spectrum of unknown compound X (Problem 7.1)

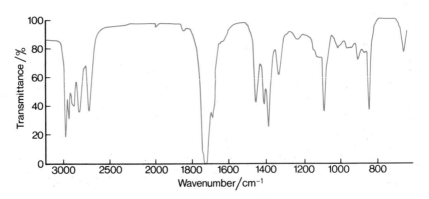

7.1b   I.r. spectrum of unknown compound X (Problem 7.1)

7.2 Identify the compound, *Y*, whose n.m.r., u.v. and i.r. spectra

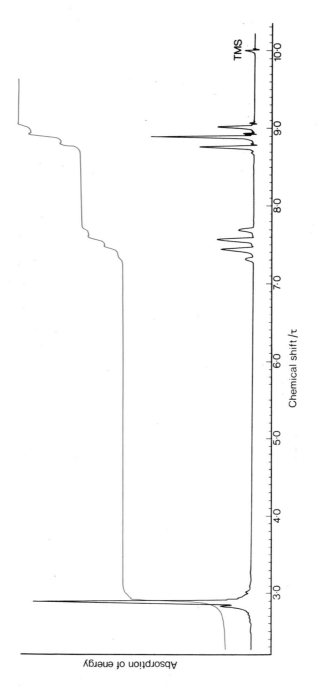

7.2a N.m.r. spectrum of unknown compound Y (Problem 7.2)

appear in Figures 7.2a–7.2c and which has the composition
90.5 % C, 9.5 % H. The most abundant peaks in the mass spectrum
of Y are as follows:

m/e	91	106	51	39	65	77	92	78	27	105
relative abundance/%	100	31	13	10	8	8	8	7	6	6

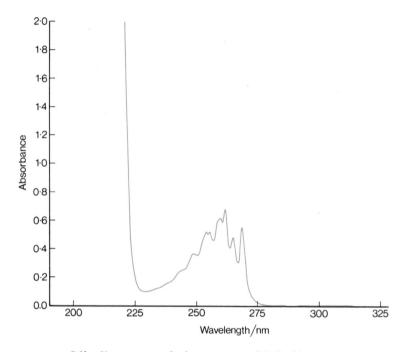

7.2b   U.v. spectrum of unknown compound Y (Problem 7.2)

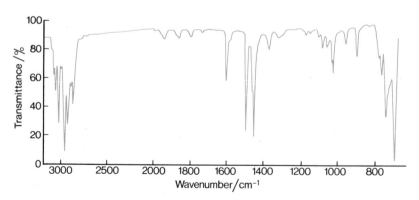

7.2c   I.r. spectrum of unknown compound Y (Problem 7.2)

Index

157